THE
PLANT LOVER'S GUIDE
TO
ASTERS

THE **PLANT LOVER'S GUIDE** TO
ASTERS

PAUL PICTON & HELEN PICTON

TIMBER PRESS
PORTLAND · LONDON

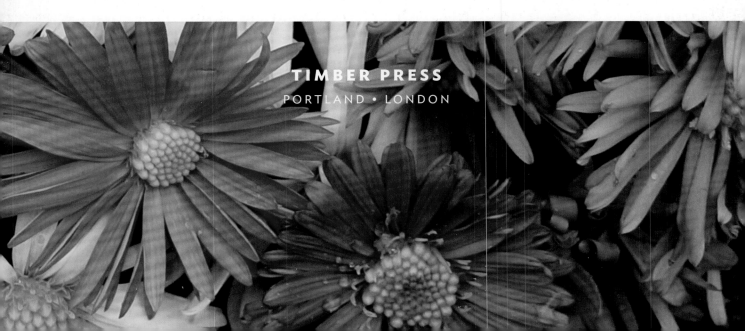

CONTENTS

75
101 Asters for the Garden

207
Growing and Propagating

WHY WE LOVE ASTERS

In this book, we hope to introduce you to the fantastic world of asters, arguably the ultimate easy-to-grow plants, providing late-season colour for plant lovers in temperate regions around the world.

Asters bring to life the garden designer's dream of soft-coloured autumnal mists, sweeping gently through the landscape.

Gardeners are undoubtedly among the luckiest people, particularly in having such a wide choice of plants readily available to allow indulgence in their passion. The large numbers of asters offered by commercial sources can make the job of choosing a few varieties to grow a lengthy, sometimes confusing, but invariably pleasurable experience.

More than 400 varieties of autumn-blooming asters grow in profusion with many other late-season perennials and shrubs at the Picton Garden.

Symphyotrichum novi-belgii
'Dusky Maid'

Whether your planting scheme is formal, more naturalistic, or even truly wild, it is easy to find asters to suit it. Varieties of *Symphyotrichum novi-belgii*, for instance, with their dense mass of flowers, lend themselves to a more formal setting, as do the large-flowered European asters where their more compact habits can be put to good use. Smaller flowered species and cultivars can be easily incorporated into naturalistic planting schemes, where a billowing cloud of flowers blurs the formal lines of borders and clumps. *Symphyotrichum* 'Little Carlow', with its generous sprays of lavender-blue flowers, is extensively used in this way; *S.* 'Ochtendgloren' is a superb counterpart with pink flowers; and *S.* 'Oktoberlicht' brings white flowers to this trio of time-tested plants. It is of no use trying to contain these varieties within trimmed hedges, as sprays will inevitably lean over and soften the stark lines.

You can achieve striking combinations by using large-flowered or bold-coloured varieties alongside the soft forms of the small-flowered asters. The glorious, deep lavender-blue flowers of *Aster amellus* 'King George' appear even bolder positioned in front of the soft creaminess of *Symphyotrichum ericoides* 'Golden Spray'. You can take a step further with certain asters that spread vigorously, *S. laeve* 'Vesta' to add a name, which will quite happily compete with ornamental grasses and other herbaceous plants in wild planting schemes, providing an eruption of colour late in the growing season.

You can also find asters to stand alone to catch the garden visitor's eye, with a beacon of colour. It is easy to visualize a group of *Symphyotrichum novae-angliae* 'Andenken an Alma Pötschke' flaming away with uniquely vibrant flower colour in an otherwise rather tired and drab late autumn border. Other varieties might be better suited to large groups forming drifts of cloud-like, star-studded sprays. A group of flowering asters will attract bees and butterflies into the garden, and later on the seedheads will be appreciated by small songbirds.

Asters, the stars of autumn, brave Mother Nature's frosty winter fingers to bring the growing season to a close with a loud, vibrant tapestry of colour rather than the murmur of mellow colour we would otherwise be left with as an introduction to winter.

A row of asters planted in the cutting border or vegetable plot will provide generous sprays of cut flowers right from the first year. Even in our over-sophisticated era, such floral riches are still gathered from gardens to decorate churches around the twenty-ninth of September, Michaelmas Day, when harvest festivals take place. Although some asters bloom in the spring and early summer, such as *Aster alpinus* and *A. coloradoensis*, the vast majority are autumn flowering. Asters, the stars of autumn, brave Mother Nature's frosty winter fingers to bring the growing season to a close with a loud, vibrant tapestry of colour rather than the murmur of mellow colour we would otherwise be left with as an introduction to winter.

Even a single clump of asters, such as *Symphyotrichum novi-belgii* 'Sarah Ballard', can give new life to a fading border.

Asters also work well within the increasing trend of growing plants in containers. Given the housing situation in the twenty-first century, plenty of enthusiastic plant lovers have tiny gardens or no real garden at all. Young asters can be used to plant up containers in the spring and will provide a colourful display in the autumn months. A handful of shoots from a friend with a garden can be potted to grow on for the same effect. Varieties of *Symphyotrichum novi-belgii* are fully at home in containers and offer the widest range of colours and heights to suit all tastes. With the right timing, they also take stopping or pinching back to produce more bushy plants.

Asters present an array of floral colours so subtle that it can be difficult to precisely describe (ok, white is not a problem!) and can sometimes be just as tricky to photograph. Although digital photography has proved to be an immensely useful tool to the publishers of plant catalogues, websites, and gardening books, even the best photography will struggle to represent the soft, foamy, must-touch appeal of a group of small-flowered asters, such as *Symphyotrichum ericoides* 'Golden Spray', or a flower colour falling somewhere between blue and violet with a hint of pink. Remember that the elusive colour tints of many aster flowers will be enhanced by the early morning and evening light when the autumnal sun is low in the sky, and will appear quite different in midday sunlight.

The cloud of stars produced by a plant such as *Symphyotrichum* 'Photograph' can be difficult to capture in a photograph, even the colours can be hard to show accurately.

Asters provide colour right up until winter begins to take a firm hold.

In this book, we have described a selection of old and new varieties that we deem exceptionally garden-worthy. Whether you are new to the gardening world or an old hand, we hope you will find useful information here on the practicalities of growing asters and using them in your garden. But before advancing any further one major point needs to be covered: What exactly are we talking about when we say "aster?"

> Whether you are new to the gardening world or an old hand, we hope you will find useful information here on the practicalities of growing asters and using them in your garden.

Up until the mid-1990s, all plants within this group were members of a single genus, *Aster*. Genetic work carried out by John Semple and his team at the University of Waterloo in Canada over a period of forty years led to a re-evaluation of the relationship of these plants and as a consequence the names used for them. Because of this work, there are now thirteen different genera in North America, including *Symphyotrichum* and *Eurybia*, but no *Aster*. However, *Aster* still holds its place as the scientific name for the European and Asiatic species and their garden hybrids. This reclassification was adopted in America early on in the scientific and commercial horticultural worlds, whereas the United Kingdom lagged behind in both fields. It is only now that we are adopting the same system as America, meaning that finally we are all using the same names.

The new naming system has made it difficult to discuss as a group the plants that bring such joy to our gardens in the late summer and autumn, despite the visual similarities. By a stroke of good fortune—or, perhaps it was sound thinking—*The Flora of North America* and official websites retain the name of aster in the form of popular or local plant names. For example, *Symphyotrichum novae-angliae* is still known as New England aster and *S. cordifolium* can be called blue wood aster. In the United Kingdom, these common names are not so familiar with the majority of autumn-flowering asters being lumped under the common name of Michaelmas daisy. However, since the name change is very recent, in this Plant Lover's Guide, we have taken the small liberty of leading with the well-known name of "Aster," working on the same basis that a book entitled "Dogs" would be more attractive to readers than one entitled "*Canis lupus familiaris.*" We hope this will allow us to be understood on both sides of the Atlantic.

DESIGNING WITH ASTERS

Even as a little girl, I (Helen) delighted in the vibrant blaze of colour that transformed our garden every autumn. I was not aware of it at the time but I was witnessing the start of a very long journey. Realization came many years later when as a young woman I stood beside my father admiring those same vibrant colours and listening to him describe how he had almost finally achieved the Monet-style picture that he had been painting for nearly thirty years. My father had used Mother Nature's autumnal palette, and looking down across the bold borders my eyes were opened to the true ethos of garden creation, and my mind was inspired.

How you choose plants for your garden and how you use them are in many ways a personal matter. The design and planting of your garden is directed by the nature of your garden, light levels, soil types, temperatures, and your personality and style. This chapter does not aim to dictate how you should use the plants, but rather to explain how others have and do use them. In the process, we hope you will be inspired to try out some new plants and new combinations, and generally have some fun with your patch no matter how large or small it might be.

For the majority of gardeners, the world of garden design, designers, and the theories that go with them are far removed from the everyday reality of creating and maintaining our gardens. However, the plants available to purchase, the gardens pictured in the magazines aiming to inspire us, and of course, the gardens we visit are all affected by the current ideas and fashions in the design world.

Modern thinking has set the scene for freeing ourselves from the bonds of tradition and fashion. As gardeners in the twenty-first century, we are not obliged to follow the footpaths of trendsetters. We will not be socially ostracized for experimenting with colour combinations, even if we choose Christopher Lloyd's infamous pink and yellow plant mixtures. Rather, we are more willing to embrace and even admire something that is new and different. However, before we can experiment, we sometimes need inspiration. So, before you dash out to try mixing your very own autumn palette with asters to create your centrepieces, it's worth looking at how they've been used historically and how others are using them now.

Painting with Mother Nature's palette in the Picton Garden.

Historic Examples

In the late nineteenth century, the renowned William Robinson led a revolution in garden design ideals. Previously design had fallen into two categories: cottage garden or formal gardens often involving grandiose bedding displays. Robinson was an advocate of naturalistic planting, the echoes of which can be seen in modern design theories. A firm follower of Robinson's ideas and an active practitioner was his friend and contemporary Gertrude Jekyll. Their more informal approach suited asters, with their light and airy growing habits and pastel colours.

The main way that Gertrude Jekyll achieved less formalized looking borders was a simple adaption of block planting, which had long prevailed as the way to plant perennials and shrubs. This method meant that groups of the same plant formed squares or diamonds, giving a formal feel to the layout and a restricted number of plant combinations for each group. Jekyll took this idea and extended the blocks to be rectangular in shape

The National Collection of *Aster amellus*, *Symphyotrichum ericoides*, and *S. cordifolium* at Upton House and Gardens, a National Trust property in Warwickshire.

At the Plant Heritage National Collection in the Picton Garden, Colwall, a modern-day example of using autumn-flowering asters in a formal border similar to how asters would have been seen in their heyday.

This National Collection of *Symphyotrichum novae-angliae* is housed at Avondale Nurseries in West Midlands, where Brian Ellis tends it.

rather than square. This allowed an increased number of changing plant combinations, depending on where the viewer was standing. This method also gave the planting schemes a much softer feel.

The commercial breeding of asters took off in the early twentieth century, and over the next fifty years huge numbers of cultivars, mainly New York asters, were produced. Advances were made in colour range, flower size, form, and height range, and this commercial boom went hand in hand with a dramatic rise in their popularity as garden plants. It was not long before asters were found in almost all gardens. Often they were used to magnificent effect in Edwardian formal borders, such as at Sandford's Nurseries of Barton Mills on the Cambridgeshire-Suffolk border.

From the end of the 1960s onwards, asters fell from popularity and became virtually unheard of in design circles. It has taken a return to herbaceous borders and the increasingly popular prairie-style planting, as well as a drive to extend seasonal interest, for a revival to begin. This would not have been possible without the hard work of the Plant Heritage National Collection holders. They have played an important role in bringing the plants to the attention of the public and by preserving cultivars and species that would otherwise have disappeared altogether. These now form the foundation for future breeding work.

Asters for Formal Borders

Aster amellus 'Forncett Flourish'
Aster amellus 'King George'
Aster amellus 'Nocturne'
Symphyotrichum ericoides 'Golden Spray'
Symphyotrichum lateriflorum 'Lady in Black'
Symphyotrichum novi-belgii 'Chequers'
Symphyotrichum novi-belgii 'Remembrance'
Symphyotrichum novi-belgii 'Saint Egwyn'
Symphyotrichum 'Ochtendgloren'

Prairie-Style Planting

A major change in landscape design occurred when the emphasis started to shift from plants simply enhancing the landscape to plants playing the main role. This is the main driving theory behind the work of the well-known and highly regarded designer Piet Oudolf. It is safe to say that he was responsible for launching the currently popular "prairie-style planting" in the United Kingdom, although his own work has now gone beyond this.

A foundation idea behind prairie-style planting is the use of informal mixed planting schemes to create a landscape. These planting schemes use not only colour but also the form of the plants to create idealized en masse displays reflecting plant combinations found in wild areas such as the prairies of North America and European or Himalayan alpine meadows. The plants are selected for their longevity, ease of maintenance, and often ability to naturalize. While this may sound like a heavenly idea, to create such a design successfully is difficult, particularly on a small scale. However, visiting one of Piet Oudolf's gardens is certainly inspirational and many of the combinations can be translated into smaller gardens.

Oudolf's designs are often heavily weighted towards grasses since they provide structure for a substantial part of the year, including the winter months. These are then mixed with other plants, usually herbaceous perennials, a few shrubs, and bulbs for spring interest. The planting schemes often look their best in midsummer and early autumn since this is when herbaceous perennials tend to come into their own.

Within Oudolf's planting schemes a few asters can be found. Appearing on a regular basis is *Symphyotrichum* 'Little Carlow' because it exhibits the longevity needed and blends in with other perennials and grasses. *Aster amellus* and *S. novae-angliae* cultivars also feature, both of which combine well in large mixed plantings.

Although prairie-style planting is best suited for a grand scale, it is possible to take inspiration from these ideas and include sections in smaller planting schemes. One specific idea is combining perennial grasses and herbaceous plants to provide structure in a garden with virtually no shrubs. Asters can play a major role when applying this design concept to smaller scale gardens.

Asters and grasses provide structure in the garden at Ragley Hall, Warwickshire, United Kingdom.

Symphyotrichum laeve 'Calliope' with *Miscanthus sinensis* 'Malepartus' and *Stipa gigantea* in the prairie-style planting at Ragley Hall, Warwickshire, United Kingdom.

This combination of fine grasses and billowing mass of *Symphyotrichum* 'Little Carlow' in John Massey's garden at Ashwood Nurseries is a good example of how to use asters and grasses.

As many of the perennial grasses flower and begin to change into their autumn colours—browns, pale golds, russets, and almost beige—they become the perfect foil for the vibrant colours of the asters. When combining them it is worth thinking about the structure of both your chosen grass and the aster you want to use. For instance, *Symphyotrichum laeve* 'Calliope' is very willowy with an open form when in flower and would perhaps not be a wise choice with a lightweight grass, but looks superb combined with the strong-coloured *Miscanthus sinensis* 'Malepartus' or the bold foliage of *M. sinensis* var. *condensatus* 'Cosmopolitan'.

The effect is reversed with those asters that produce a dense mass of flowers, such as *Symphyotrichum* 'Little Carlow'. Combining these with heavyweight grasses can fail to bring the plant to the forefront; it is often better to use finer grasses with an upright or arching habit.

Sometimes rather than using complementary but contrasting forms to create our combinations, it is a good idea to use plants that reflect each other's forms. This can be achieved through structural similarities, like the clasping leaves and fine arching sprays of *Symphyotrichum turbinellum* that resemble many of the lighter grasses. Colour, too, can create a wistful harmonious combination as, for instance, when using pale pink asters, such as *Aster amellus* 'Rosa Erfüllung', with the bleached beige seedheads of Mexican feather grass (*Stipa tenuissima*) or something similar.

Light and airy *Symphyotrichum turbinellum* reflects the refined structure of nearby grasses.

The seedheads of *Doellingeria umbellata* are as attractive as the flowers and last through the winter.

Another important concept to emerge during recent years is that the stems and seedheads of the plants once flowering has finished are not an unsightly mess, but are in fact excellent for lengthening the season of interest in the garden and keeping structure in the garden for a significant portion of the winter. Here again many asters can have a starring role. A prime example is *Doellingeria umbellata*, which flowers in late summer and then produces a mass of silvery seedheads. As these blow away, the umbrella-like structure of the spray is left to catch the frosts and gradually to become bleached as the winter advances.

It is worth considering the structure of neighbouring plants to get the most from their winter remains. For instance, Joe Pye weed (*Eutrochium purpureum*) has big dense seedheads, whereas many of the taller willowy asters have a mass of small seedheads; when planted near each other, these can provide an interesting contrast. The idea of leaving the old stems on until late winter does not appeal to everyone, but an added bonus is that the seeds are also a good source of food for birds over winter and an excellent sanctuary for insects.

Finally and most importantly for asters, prairie-style planting has served to revitalize the interest in using herbaceous perennials in the garden, whether this be in large groups or a series of specimen plants. Herbaceous borders are seeing a revival in their fortunes albeit in altered forms from the late nineteenth and early twentieth century.

Asters for Naturalistic Plantings

Aster amellus 'Forncett Flourish'
Aster amellus 'King George'
Doellingeria umbellata
Symphyotrichum ericoides 'Golden Spray'
Symphyotrichum laeve 'Calliope'
Symphyotrichum 'Les Moutiers'
Symphyotrichum novae-angliae 'Primrose Upward'
Symphyotrichum novi-belgii 'Elta'
Symphyotrichum novi-belgii 'Goliath'
Symphyotrichum novi-belgii 'Beechwood Charm'
Symphyotrichum 'Ochtendgloren'
Symphyotrichum 'Prairie Purple'

Herbaceous Borders

Asters have long played an important role in herbaceous borders providing colour at vital points, their main role being late season once the major players of high summer have faded. They are among the specialist performers of the plant world. Competent in a wide variety of shows, these adaptable plants can do the extravagant, spectacular, brilliant hit and just as easily calm down to enchanting, simple, and gentle. Choices can be made from a cast of thousands, if the stage is big enough, down to a one-man show. Sometimes it is enough to have only one or two specimens of asters in the garden and enjoy them for being just what they are, providing that final lift before winter.

The most traditional form of herbaceous border, and the one that springs to mind when the term is mentioned, is the long straight or curved border with groups of plants used along the length to create a riot of colour from spring until late autumn. Often groups of the same plant or same colour are repeated at regular intervals to draw the eye into the depths of the border.

A single specimen of *Symphyotrichum novi-belgii* adds colour and interest to an herbaceous border in early autumn at Rob and Diane Cole's garden, Meadow Farm, in Feckenham, Worcestershire.

The long border at Waterperry, Oxfordshire, is a brilliant example of the traditional herbaceous border with the repetition of colour. At the back golden yellow arching sprays of *Solidago canadensis* and the globular heads of *Eutrochium purpureum* complement the shades of lavender, pink, and purple, which dominate in early autumn.

Asters as Accent Plants

Aster amellus 'Forncett Flourish'
Aster amellus 'Rosa Erfüllung'
Aster ×frikartii 'Wunder von Stäfa'
Symphyotrichum 'Anja's Choice'
Symphyotrichum novi-belgii 'Freya'
Symphyotrichum novi-belgii 'Heinz Richard'
Symphyotrichum oblongifolium 'Fanny's Aster'
Symphyotrichum turbinellum

Here in the herbaceous border many asters can be shown to their best. The width of the border and the length affect which asters are chosen and how many are planted in each group. Low-growing varieties, such as the deep pink diminutive *Symphyotrichum novi-belgii* 'Rosenwichtel' or the charming lavender-blue 'Lady in Blue', are best used at the front of the border, whereas those statuesque players who need a little more elbow room, such as *S. novae-angliae* 'Purple Cloud' or *S. laeve* 'Nightshade', will find a perfect home at the back. However, varieties with an open habit can also be used at the front despite being tall. *Aster ×frikartii*, for instance, will provide a hazy dome of lavender-blue from midsummer until the frosts arrive.

In the 1950s, Alan Bloom showed the world a different sort of herbaceous planting with his curving island beds at Bressingham. These magnificent borders contained plenty of asters, adding colour from late summer until late autumn, and soon became very popular additions to the English garden. With this type of border, taller plants are generally positioned in the centre of the bed.

It would be an oversight not to mention Christopher Lloyd at this point. He transformed his garden at Great Dixter into an Eden of intensively planted borders using the Arts and Crafts cottage garden style. He was not inhibited by popular opinion—colours slopped off the palette into the border before any one could say, "That will not look good together." As a result, Great Dixter has become famous for this luxurious and uninhibited style of

Asters for Cottage Gardens

Aster amellus 'King George'
Aster amellus 'Nocturne'
Symphyotrichum 'Les Moutiers'
Symphyotrichum novae-angliae 'Barr's Pink'
Symphyotrichum novi-belgii 'Sam Banham'

planting, which looks so easy but can take so much time and care to achieve the length of flowering season and constantly vibrant living displays. Often this length of flowering is achieved by growing plants in pots, which are then put out in the border around groups of permanent planting that have faded or are yet to come. Asters are well suited to playing a role in this form of gardening.

Plants are graded in height towards a tall central point in the island border at Waterperry, Oxfordshire. *Aster pyrenaeus* 'Lutetia' has been used at the front in the centre, with the taller *Symphyotrichum* 'Little Carlow' on the left and *S. novae-angliae* 'Helen Picton' on the right further back.

Of course, not all of us have large gardens, but that does not mean we cannot take ideas and try them on a smaller scale. In the less formal herbaceous border, asters are quite at home. You can achieve exciting combinations with relatively few plants and there is no need to restrict yourself to perennials alone either. Tender perennials and annuals used alongside herbaceous perennials and shrubs can create the most stunning plantings. I (Helen) have fallen in love with the combination, created rather by accident, of

ornamental tobacco (*Nicotiana mutabilis*) towards the back of the herbaceous border and *Symphyotrichum novi-belgii* 'Elta' in front of it. The nicotiana thrives in the lightly shaded position, producing a mass of white and pale purple-pink flowers over large mid-green foliage throughout the summer and autumn. The aster goes virtually unnoticed in front of this magnificent plant until it erupts into dense sprays of bright purple-pink, whereupon the combination becomes a graceful harmony of pink and white.

The much-neglected tender perennials include cannas, bananas, and dahlias. Much like asters, dahlias suffered a fall from grace having previously been extremely popular, and now their usefulness as a garden plant and as a cut flower is being appreciated once again. However, this time the planting trend has moved away from rows of dahlias or entire beds of only dahlias towards a happier arrangement whereby they share border space with all the other plants. Here is where two plant groups that have suffered through the fickle turns of fashion only to flourish again join forces to show the world what the word *colour* really means.

When choosing dahlias to plant with asters, consider flower form and colour, and foliage colour. For instance, the bright red decorative *Dahlia* 'Witteman's Best' looks superb against the smaller deep lavender-blue of *Symphyotrichum novi-belgii* 'Marie's

Vibrant colour and form combinations can be created using dahlias and asters. Here, *Symphyotrichum novi-belgii* 'Marie's Pretty Please' and *Chrysanthemum* 'Belle' make a lovely display with *Dahlia* 'Witteman's Best'.

Rhus typhina can be used to create stunning autumnal displays with a range of asters and other late-season perennials such as *Rudbeckia subtomentosa*.

Symphyotrichum novi-belgii 'Sarah Ballard' growing with *Bergenia*, *Lamium*, and *Hosta* in front of the rich purple-red foliage of *Cercis canadensis* 'Forest Pansy' brought this area to life in the autumn when it may otherwise have been easy to overlook.

Pretty Please', whereas the dark-leaved *D.* 'Twyning's After Eight' is better suited as a foil for the small-flowered *S. ericoides* 'Pink Cloud' whose pale pink puffs stand out in glorious denial of approaching winter against the rich background of the dahlia leaves while large single cream flowers watch from above.

Shrubs are another useful—some may say essential—addition to the herbaceous border that are all too often neglected. A friend recently told me that she had "been afraid of using shrubs in her garden in case it was a waste of space." This sentiment is all too often repeated. Shrubs when chosen well can add interest at all points of the year and give a border permanent structure. They often have beautiful flowers and usually have either coloured foliage throughout the year or good autumn colour. Even the evergreens have a role to play as structural must-haves and as an excellent background and foil for multitudes of flowers throughout the summer and autumn.

One of the shrubs or rather small trees that does well in our Colwall garden is the stag's horn sumach, *Rhus typhina*. Yes, it can sucker, but the answer is not to cut the top off it. With its wonderful architectural furry stems and serrated foliage it is pretty in both winter and summer; however, its pièce de résistance is the beautiful mellow yellows, oranges, and burnished reds of the autumn foliage. A better shrub to work with grasses and late-season herbaceous perennials, including *Symphyotrichum novae-angliae* and *S. laeve* cultivars, is hard to imagine.

We have also used a wide range of Japanese maples and conifers all of which can add that extra structure, texture, and interest to the garden. A more surprising combination and a most effective one if, unlike us, you are lucky enough to be able to grow *Cercis canadensis* 'Forest Pansy' was seen at Meadow Farm in Worcestershire.

Staying with shrubs brings us to an important point about cottage garden style plantings. Borders packed with beautiful flowers overflowing onto meandering paths or lawns, with floriferous shrubs such as hydrangeas or roses here and there, are all well and good. However, without any solid colour or form there is a danger that the billowing mass, particularly when it comes to large numbers of asters, can just be lost into the background. A low-trimmed hedge, simple topiary ball, or just a dense, strong, plain-coloured shrub can help bring them all back into focus.

Olive Mason's modern take on the cottage garden with masses of *Symphyotrichum* 'Les Moutiers' at Dial Park, Worcestershire.

The smooth formal lines of the trimmed variegated box (*Buxus sempervirens* 'Elegantissima') are the perfect background for the sea of lavender *Aster amellus*.

The rose garden at Ragley Hall, Warwickshire, where the roses are planted alongside asters and other herbaceous plants, trees, shrubs, and bulbs.

Roses Need Friends and Other Unexpected Places for Asters

Away from the traditional positions such as herbaceous borders, asters can find homes in places that may seem surprising to the uninitiated. An increasingly common idea is that to reduce the need for chemicals in the garden, and to keep roses healthy and happy, they need to be planted with other plants. The increased diversity encourages more wildlife to the garden including the natural predators of pesky visitors, namely, aphids.

A more truthful sentiment than "roses need friends" has rarely been uttered, but how does this relate directly to asters? The answer is twofold. Firstly, many roses will have a second flush of flowers, and some are even naturally late, which means that either way their flowering coincides with the earliest of the autumn asters. For example, a stunning display can be achieved using the lavender-blue mounds of *Aster ×frikartii* 'Wunder von Stäfa' to complement the deep pink *Rosa* 'Braveheart'. Secondly, if the roses do not have a second flush, a rose garden can be empty of colour and interest for a remarkably long time. This can be solved by using asters to bring the garden back to life for the autumn without detracting from the earlier beauty of the roses.

The informal in the formal: *Symphyotrichum novi-belgii* 'Little Man in Blue' and other asters add late-season interest to the rose garden at Ragley Hall, Warwickshire.

Symphyotrichum 'Prairie Purple' in the Rose Garden at Ragley Hall, Warwickshire.

IN THE ROCK GARDEN

The rock garden may also seem like an unlikely place for asters. However, even if you are a stickler for only alpines being allowed in a rock garden, you'll find asters that are not only suitable but almost require the free draining and hard conditions to thrive. *Eurybia sibirica*, a true alpine no matter what definition you use, has a very compact nature and is hardy to the extreme but does not appreciate being too wet at the root. Planted on the lower portions of the rock garden it makes beautiful domes of flower in late summer. One of the most effective positions though is probably in the crevice garden, again on lower portions, where this alpine aster will fill a crevice, flowering along the entire length given time.

Other prime choices for the rock garden or container are the various cultivars of *Aster alpinus* with their beautiful late spring daisies providing a lift when many other alpines have begun to fade. Although slightly taller than *A. alpinus* cultivars at 30 cm (12 in.), *A. thomsonii* 'Nanus' can still be used in the rock garden without being out of scale and brings a welcome haze of lavender-blue from midsummer until midautumn.

Eurybia sibirica growing on a crevice garden.

Aster thomsonii 'Nanus' tops the favourites list of truly dwarf asters for summer to fall colour on rock gardens or raised beds. It truly looks at home in these situations.

It is worth bearing in mind that numerous species of "true" asters, native to Europe and Asia, inhabit upland and mountain areas. Growing conditions are constrained by altitude, light levels, winter rest periods, spring moisture supplies, and the all-important soil structure and critical balance nutrients. This can explain why some of these species and their hybrids are unable to flourish in warmer or more extreme climatic zones. The superb *Aster ×frikartii*, almost universally lauded in northern Europe, has proved to be less than successful in many parts of North America. In light of this, perhaps *A. amellus* and its relatives are worth thinking about for late-season colour in a rock garden or raised bed. A little cast shade from a rock or bush can cool down the midday heat. More grit in the planting mix will encourage less growth if full-sized plants might look out of scale. Try growing some plants in scree mixes or rock crevices to reduce the plants still further and achieve a sparse, natural effect. We have achieved good results from plants growing in tufa crevices.

Asters for Rock Gardens

Aster alpinus and cultivars
Aster diplostephioides
Aster ×frikartii 'Flora's Delight'
Aster thomsonii 'Nanus'
Aster tongolensis 'Napsbury'
Eurybia sibirica
Symphyotrichum novi-belgii 'Rosenwichtel'

Symphyotrichum ericoides f. *prostratum* 'Snow Flurry' can make an exciting container plant providing interest in midautumn.

Asters for Containers

Aster alpinus cultivars
Aster ×frikartii 'Flora's Delight'
Aster tongolensis 'Napsbury'
Symphyotrichum 'Coombe Fishacre'
Symphyotrichum ericoides 'Golden Spray'
Symphyotrichum ericoides f. *prostratum* 'Snow Flurry'
Symphyotrichum lateriflorum 'Prince'
Symphyotrichum novi-belgii 'Apollo'
Symphyotrichum novi-belgii 'Dietgard'
Symphyotrichum novi-belgii 'Lady in Blue'
Symphyotrichum novi-belgii 'Rosebud'
Symphyotrichum 'Ochtendgloren'

AS CONTAINER PLANTS

Asters can make attractive container plants as long as they are carefully selected with certain criteria in mind. A single plant can be effective in a container, particularly if it has a very bushy habit like that of *Symphyotrichum* 'Coombe Fishacre'. Often a longer-lasting container can be created using asters in the centre with lower-growing summer-flowering plants including annuals around them.

A highly effective aster for containers, *Symphyotrichum ericoides* f. *prostratum* 'Snow Flurry' easily fills a large shallow container and spills over the edges. The form is interesting enough to warrant the space even when it is not in flower and the mass of tiny white flowers on the low growth can be quite breathtaking. In particularly large containers, small-growing grasses can be used with asters to create a miniaturized garden. This pairing will provide interest and colour throughout the growing season, even well into winter when the structure can be appreciated in much the same way as it is in the garden on a larger scale.

At Waterperry, the stock plants are rowed out much as you would do in a vegetable garden or allotment.

IN THE VEGETABLE GARDEN

Although in times past asters were grown in the vegetable garden or allotment, such treatment is not nearly as popular as it once was. When grown in the vegetable garden, asters can be treated much like vegetables or other cut flowers, rowed out and staked if necessary. Then the flowers can be harvested when they bloom, providing fabulous bouquets for the house or the enterprising market gardener.

Many of the asters available today were bred with use as cut flowers in the home in mind. This is particularly evident in the range of flower size, spray strength, and shape found within the *Symphyotrichum novi-belgii* cultivars. However, *Aster amellus* and its cultivars make excellent cut flowers provided they have reasonable stem length, as do *A. ×frikartii* and its cultivars; in fact, almost all asters can be used for cutting. The only group to be careful with is *S. novae-angliae*. They need to be picked the night before use, on a good long stem and soaked in a deep bucket of water with either flower preservative or a mix of aspirin and sugar to stop them from closing up. Even then, they tend to close when it becomes dark.

This large arrangement features *Symphyotrichum novi-belgii* and small-flowered aster cultivars.

Asters for Cut Flowers

Aster amellus 'Gründer'
Aster amellus 'Nocturne'
Aster ×*frikartii*
Aster tongolensis 'Napsbury'
Symphyotrichum ericoides 'Golden Spray'
Symphyotrichum 'Kylie'
Symphyotrichum novi-belgii cultivars
Symphyotrichum 'Oktoberlicht'

Sprays of mixed *Symphyotrichum novi-belgii* are also ideal for smaller flower arrangements.

Because of its relatively long, straight stem, the large lilac flower of *Aster amellus* 'Nocturne' is ideal as a cut flower.

COMPANION PLANTS

You can look at companion plants for asters in a number of ways. The first is companion plants for early-season colour, for which spring bulbs such as daffodils, snowdrops, and tulips play a useful role. They will put on a wonderful show in the early part of the year while the foliage of the asters is barely showing, and by the time the asters have grown the bulbs will have finished with their foliage. Later in the year, asters can be planted with summer herbaceous perennials and shrubs, which will provide interest before the asters come into flower.

To create a mass display of autumn colour you can combine asters with other autumn-flowering perennials, bulbs, grasses, and autumn-colouring shrubs. For this dream boost of colour, a few plants that can be considered staples of the autumnal planting scheme are *Rudbeckia fulgida* var. *deamii*, *R. subtomentosa*, *Echinacea purpurea*, *Helenium* 'Sahin's Early Flowerer', *Solidago rugosa* 'Fireworks', *Aconitum carmichaelii* 'Arendsii', and *Persicaria amplexicaulis* 'Orange Field'.

An option that shouldn't be overlooked is simply using asters as companions for other asters. This way you can create that real "wow" factor late in the season and appreciate the simple beauty and charm of this group of plants. Much in the same way as when working with other companion plants, great combinations can be created through colour and through

Rudbeckia fulgida var. *deamii* is a tough easy-going plant for a sunny location reaching a height of about 90 cm (36 in.). It has a long summer and autumn flowering season and its rich golden flowers mix well with asters of any colour, purple being the most striking of all.

The purple flowers of *Symphyotrichum novae-angliae* 'Helen Picton' combine well with a pink Japanese anemone.

You can highlight the large bold flowers of *Aster amellus* 'King George' by planting it near the wonderfully tough and long flowering *Persicaria amplexicaulis* 'Orange Field' with its soft, drooping flower spikes.

However you choose to use asters in your garden, there is an aster for almost any position.

The autumn garden explodes with colour when the daisy flowers of *Symphyotrichum novi-belgii* cultivars appear.

different forms. One of our favourite combinations is the large deep lavender-blue flowered *Aster amellus* 'King George' used in front of the soft form and colour of *Symphyotrichum ericoides* 'Golden Spray'. Another combination that we use in a number of places around the garden is *S.* 'Little Carlow' alongside *S. novae-angliae* cultivars; the dense sprays of flowers bring out the best in the New England asters without making them look scruffy.

Some aster colour combinations work so well that we tend to use them repeatedly. The first is the fluffy white flowers of *Symphyotrichum novi-belgii* 'Sandford's White Swan' with the deep rich purple flowers of *S. novi-belgii* 'Purple Dome'. Secondly, and on a slightly more vibrant note, is *S. novi-belgii* 'Faith' with its large purple-blue flowers against a purple-pink variety such as *S. novi-belgii* 'Elta'. However, the top of our combination list would have to be *S. novi-belgii* 'Lassie' with 'Blauglut' planted just in front of it. Not only do the colours of these two asters perfectly complement each other, but also they both have a longer-than-average flowering season making them ideal companions.

A red admiral butterfly visits a
Symphyotrichum novae-angliae plant.

A comma butterfly looks for nectar.

It is also important to think about the benefits that mixed plantings can bring to the wildlife in your area. Asters have always been considered excellent for pollinating insects, especially bees, hoverflies (or syrphid flies), and butterflies. Seeing a clump of asters covered in beautiful butterflies in early autumn, or hearing the sultry buzzing of bees and hover-flies over a bed of *Aster amellus* in late summer tends to form a nostalgic and idyllic image of autumn that is hard to forget.

However you choose to use asters in your garden, there is an aster for almost any position in the garden, where their most important role will be to enliven the late summer and autumn with their cheery colourful flowers, fending off the grey of winter for a while longer.

Asters That Attract Wildlife

Aster amellus 'Gründer'
Eurybia divaricata 'Eastern Star'
Symphyotrichum 'Coombe Fishacre'
Symphyotrichum lateriflorum 'Prince'
Symphyotrichum novae-angliae 'Barr's Pink'
Symphyotrichum novae-angliae 'Mrs S T Wright'

UNDERSTANDING ASTERS

T his chapter will help give you some guidance as to which plants are likely to be best suited to your garden, disease problems you may face, general guidelines for propagation, and the range of plants found within the aster group. It also tells the story of aster breeding and how so many plants came into cultivation.

The Range of Asters

Because asters, which used to belong to a single genus, are now separated by botanists into many different genera and even more species, it is useful to have an overview of the groups mentioned in this book. Remember, in the new taxonomy, the European species remain in genus *Aster*, but North American species have been moved into new genera, three of which are described in this book: *Doellingeria*, *Eurybia*, and *Symphyotrichum*.

Regardless of how botanists classify asters, all the plants share a certain number of morphological features, the most obvious of which is the structure of the flower. The "daisy flower" is an iconic simple form that ties all members of the Asteraceae (daisy family) together; however, we are not really looking at one single flower but a head of numerous small flowers, known as florets. Within a single flower head, there are usually two different forms of these florets: disc florets, which are most often yellow and make up the eye or disc of the flower, and ray florets. Ray florets are probably the most important to gardeners as they have extended ligules, which gardeners recognize as petals. The variation in colour, form, and number of ray florets is what gives such a wonderful range of cultivars. The form of the aster flower is also hugely important in their usefulness to insects, since each "flower" is actually a one-stop shop of numerous nectar sources for insects.

An aster flower head is actually composed of many small flowers. The minute tubular flowers at the centre of the head are known as disc florets. The petal-like flowers surrounding them are the ray florets.

Aster diplostephioides, from the Himalayas, is a creeping plant with very narrow lilac-coloured ray florets around a golden centre.

Aster pyrenaeus 'Lutetia', a French selection of a rare species from the East and West Pyrenees.

EUROPEAN AND ASIATIC ASTERS

This group includes *Aster alpinus* (alpine aster), *A. amellus* (European Michaelmas daisy), *A. diplostephioides*, *A. ×frikartii*, *A. peduncularis*, *A. pyrenaeus*, *A. thomsonii*, and *A. trinervius*.

All plants within the group grow 30–120 cm (1–4 ft.) tall. They have some degree of hairy foliage, which is usually much softer than that of New England asters, and all have single flowers that are often large. With the exception of *Aster trinervius*, European and Asiatic asters flower earlier in the year than the majority of North American asters. *Aster alpinus* flowers in the spring, followed by *A. amellus* in late summer, and finally *A. trinervius*, which in some cases does not start to flower before late autumn.

Asters in this group mostly require an open sunny position with excellent winter drainage. The species native to Europe prefer an alkaline soil. Those from the Himalayas, such as *Aster peduncularis*, are more forgiving on both soil pH and winter drainage.

European and Asiatic asters are resistant to powdery mildew but slugs and snails can be a problem in the early spring. Verticillium wilt can also affect *Aster amellus* when replanted in the same soil after division. Divide this group of asters every three to five years to keep plants healthy.

Aster ×frikartii 'Wunder von Stäfa', one of many selections produced by Swiss nurseryman Carl L. Frikart, who successfully crossed European *Aster amellus* with Himalayan *A. thomsonii*.

NEW YORK ASTERS

Symphyotrichum novi-belgii (syn. *Aster novi-belgii*), the only member of this group, has been subject to the most intensive commercial breeding of all the asters. Although treated as a botanical species, it is really a group of hybrids which all share very similar features. The influences from various other species, including *S. laeve* (smooth aster) and *S. dumosum* (rice button aster), have led to a vast range in flower colour, flower form, height, disease resistance, and general garden worthiness. At the peak of their popularity in the United Kingdom, more than a thousand different cultivars were available. Now, the number is closer to three hundred, which is still a vast amount of plants from which to choose.

New York asters grow 25–180 cm (10–72 in.) tall. Their smooth foliage ranges from mid-green to purple tinted, and is usually long, pointed, and broader towards the centre of the leaf rather than heart shaped. The flowers vary tremendously in size and shape, but are always held in sprays rather than individually.

A sea of colour created by New York asters, *Symphyotrichum novi-belgii,* is hard to surpass in autumn.

Symphyotrichum novi-belgii 'Algar's Pride' produces large, single, lavender-blue flowers on tall stems.

Symphyotrichum novi-belgii 'Jenny' is probably the most intensely coloured dwarf New York aster. The bright purple-red, double flowers are large for the compact habit of the plant.

A good moisture-retentive garden soil is ideal and an open sunny position is necessary to ensure good flowering. Some New York asters will start to flower in early autumn and from there a range of cultivars fills all the gaps until late autumn.

Most New York asters are susceptible to powdery mildew, although those with *Symphyotrichum laeve* in their breeding show more resistance. They are also susceptible to Verticillium wilt particularly when re-planted for a number of years in the same ground. Regular division is very important to ensure that the plants maintain their vigour, flower size, and quantity. This can be done annually with up to a maximum of three years between divisions.

The small violet flowers of *Symphyotrichum novi-belgii* 'Jean' are quite weatherproof, standing up well to endlessly wet days in autumn.

NEW ENGLAND ASTERS

Symphyotrichum novae-angliae (syn. *Aster novae-angliae*) is the only species in this group. Despite being introduced to the United Kingdom at the same time as the New York asters (1710), New England asters have historically never elicited the same levels of interest or commercial breeding work. This is changing. At the peak of interest in asters, there were around seventy different cultivars, which then fell to about fifty. But the good news is the number has rebounded back to more than seventy and more cultivars are arriving on the scene almost every year.

New England asters mostly reach 90–200 cm (3–6.5 ft.) tall, with the exception of 'Purple Dome' and 'Vibrant Dome' which are 30 cm (12 in.) tall. The exact height of each cultivar depends on the plant's growing conditions. If well fed with plenty of moisture, plants will reach the top range of their height; poor soil or dry conditions result in a reduction in height.

This group has rough, hairy mid-green to pale green foliage. The plants grow in woody clumps, and the stems are usually self-supporting after the first year. The flowers have varying numbers of petals but the disc is always visible. Flowering is from mid to late autumn. To achieve a good show of flowers, an open sunny position is essential. The plants are not particular, growing well in any reasonable garden soil.

This group is generally very disease resistant. Very rarely a minor infection of powdery mildew will occur, but this is never serious. Rust can affect them but again this is rare. If untreated, however, rust can prove fatal. It is advisable to divide plants every three to five years, but they can often be left longer.

Symphyotrichum novae-angliae 'Colwall Constellation', one of several medium tall selections bred at Old Court Nurseries in Colwall.

The fabulous double flowers of *Symphyotrichum novi-belgii* 'Sandford's White Swan' open a soft pure white, becoming tinged with purple-pink as they mature.

When used alongside the rich yellows of many late-season herbaceous plants, such as *Rudbeckia fulgida* var. *deamii*, New England asters create a truly autumnal scene.

Wood asters can take a certain amount of light shade although in most cases the flowers will be produced more freely if grown in an open position.

WOOD ASTERS

This is a group of species that mostly all share the common name of wood aster as well as similar growing requirements and identification features. Included here are blue wood aster, *Symphyotrichum cordifolium* (syn. *Aster cordifolius*); white wood aster, *Eurybia divaricata* (syn. *A. divaricatus*); Schreber's aster, *E. schreberi* (syn. *A. schreberi*); large-leaf wood aster, *E. macrophylla* (syn. *A. macrophyllus*) and *E.* ×*herveyi*; and rough wood aster, *E. radula* (syn. *A. radula*).

Wood asters grow to 60–180 cm (2–6 ft.) and have large roughly heart-shaped foliage, which is to some degree hairy. As the common name suggests, they can take a certain amount of light shade although in most cases the flowers will be produced more freely if grown in an open position. The latter is particularly true when dealing with hybrids of *Symphyotrichum cordifolium* such as 'Little Carlow'. Any good garden soil that does not dry out too much over summer or become waterlogged over winter is ideal. The flowers are single and relatively small from 1.3 to 3.8 cm (0.5 to 1.5 in.) across in sprays. Some species are early flowering, from late summer; others don't start to flower until late autumn.

Although the species are very disease resistant, powdery mildew can sometimes infect hybrids, but it is rarely a major problem. Slugs and snails can damage the young growth but again rarely to any long-term detrimental effect. Divide plants every three to five years. In some cases, they can be left longer, particularly if used in a more wild setting.

Eurybia ×*herveyi* 'Twilight', a hybrid of large-leaf wood aster, creates a mass of colour in late summer.

HEATH ASTERS

This group includes the white heath aster, *Symphyotrichum ericoides* (syn. *Aster ericoides*); Pringle's aster, *S. pilosum* var. *pringlei* (syn. *A. pringlei*); and hybrids of both. Also included here is the hardy aromatic aster, *S. oblongifolium* (syn. *A. oblongifolius*).

Heath asters have fine heather-like foliage or linear foliage, varying from a few hairs on the midribs to hairs on the entire surface of the foliage. Flowers are small, from 1.3 to 2.5 cm (0.5 to 1 in.) across, held in bushy sprays. In general, they require an open sunny position in any good garden soil that does not become waterlogged during winter.

The plants closest to *Symphyotrichum ericoides* have the highest resistance to powdery mildew. Hybrids, particularly those closely related to *S. pilosum* var. *pringlei*, can be susceptible to powdery mildew but it is not usually a serious problem. Slug and snail damage can be serious on the young growth in the early spring. Divide plants every three to five years.

The gorgeous warm shades of autumn are punctuated by the cooler lavender-blue cultivars of *Symphyotrichum oblongifolium*, the hardy aromatic aster.

Vigorously spreading *Symphyotrichum laeve* 'Vesta' has masses of pure white flowers on dark willowy stems.

The pale lavender-blue flowers of *Symphyotrichum turbinellum* are borne from long side branches of the sprays.

OTHER NORTH AMERICAN SPECIES

In this large and varied group, the best known and most widely grown members include smooth aster, *Symphyotrichum laeve* (syn. *Aster laevis*); starved aster, *S. lateriflorum* (syn. *A. lateriflorus*); and prairie aster, *S. turbinellum*. These asters are largely free of disease, although powdery mildew occasionally appears in very hard growing years. Divide plants every three to five years.

The violet-blue starry flower of *Aster amellus* adds colour to the late summer garden.

The Story of Asters in Cultivation

The earliest record of asters in cultivation in the United Kingdom was 1596. At this time, the European aster, *Aster amellus*, was being grown in John Gerard's Holborn Physic Garden. Of course, the aster was not grown because it was a beautiful garden plant but rather because of its use in herbal medicine of the time. A tincture of aster root was said to cure a person of ravings, and a lotion made with aster root was applied to leg ulcers and varicose veins.

Once plant hunters reached the Americas, a flood of plant material began to enter English gardens and nurseries including a number of *Aster* species. One of the first was the small white-flowered Tradescant's aster (*A. tradescantii*), which was introduced to the United Kingdom in the late 1600s. Among this early rush arrived the two species that have been the most important commercially: the New England aster (*Symphyotrichum novae-angliae*) and the New York aster (*S. novi-belgii*).

From that point on there was a lull in aster interest until the late nineteenth century when Michaelmas daisy borders became fashionable. During the renovation of his garden at Aldenham in Hertfordshire, United Kingdom, the Hon. Vicary Gibbs created a Michaelmas daisy border that was more than 150 m (490 ft.) long by up to 15 m (49 ft.) wide in places. This border was planted with as many different *Aster* species as were available at the time. This rapidly led to a great deal of hybridization and the appearance of large numbers of seedlings. Rather than throwing all of these seedlings out the Hon. Vicary Gibbs and his head gardener, Edwin Beckett, began to select forms that were an improvement on the parents. They were particularly keen on improved colour intensity and flower size, since at the time the majority of asters were rather small flowered in pale shades of lavender-blue, pink, and white.

Famed garden designer Gertrude Jekyll further championed asters by advocating a Michaelmas daisy border as an essential feature for

Symphyotrichum novi-belgii 'Beauty of Colwall', one of Ernest Ballard's earliest cultivars.

any halfway decent garden. However, asters still did not really catch on as the must-have garden plant until British nurseryman Ernest Ballard began what became a lifelong obsession and commercial enterprise: the breeding of asters. It all started in the vegetable garden of his home, Old Court, where he began to row out aster seedlings. Each year he selected a few plants to go back into the breeding programme. As time passed, more and more space was being given over to these aster seedlings, much to the dissatisfaction of Ernest's vegetable gardener. Luckily, in 1906 just before the crisis point was finally reached, a nearby plot of land came up for sale and was rapidly purchased. This is where Ballard founded Old Court Nurseries and began taking a much more commercial approach to breeding.

In 1907, the RHS (Royal Horticultural Society) held a trial of asters in which more than three hundred varieties were entered. The only plant awarded a first class certificate was Ernest Ballard's *Symphyotrichum novi-belgii* 'Beauty of Colwall'. At the time, it was considered the best double form available and a grand advancement in the breeding of this genus.

Among the many aims of Ernest's breeding, the first was to increase the number of ray florets (petals) in the flower. This began with 'Beauty of Colwall' and ended in the early 1950s with 'Marie Ballard', which has up to 250 ray florets creating a wonderful formal double flower. His work also concentrated on creating pure shades of blue, red, and pink, as well as a good white. Over the years, he transformed the New York aster from small pale-coloured things into the magnificent vibrant plants that we recognize today. Other breeders and nurseries often brought out refinements of Ballard's cultivars, cleaner shades and better growth habits, but rarely did anyone beat him in the introduction of truly new advancements in the breeding.

One area in which Ernest Ballard had no interest was the breeding of dwarf asters, which did not exist before 1920. There was, however, demand among the gardening public for such plants, including one Victor Vokes, an official of the War Graves Commission, who was searching for low-growing perennials to provide autumn colour for the cemeteries in his charge. Rather than do nothing he set about breeding a range of dwarf asters by hybridizing the diminutive but rather dull rice button

aster (*Symphyotrichum dumosum*) and the brightly coloured New York aster (*S. novi-belgii*). After four years, Vokes had his first set of viable seed, with his first dwarf hybrids coming into flower in 1925, at heights between 20 and 45 cm (8–18 in.). A further seven years saw this new race of asters tested in various soils to make sure that they stayed true to form and flowered freely throughout the vagaries of the English autumn. Many German growers have continued this work, including Heinz Klose who has raised numerous fine cultivars since the 1960s.

The successful crossing of European *Aster amellus* with Himalayan *A. thomsonii* by Swiss nurseryman Carl Ludwig Frikart around 1918 was an important breakthrough in breeding asters. This hybrid had appeared once before at a British flower show in 1892, but thereafter not seen until Frikart introduced three varieties—'Eiger', 'Jungfrau', and 'Mönch'—named for the three peaks visible from his nursery. *Aster ×frikartii* 'Mönch' quickly became a worldwide gardening triumph with its gorgeous, large lavender-blue flowers and extended flowering season from midsummer to midautumn, plus hybrid vigour at root level.

Meanwhile the race among breeders to bring out new and innovative cultivars escalated. Every year these new plants could be seen at the flower shows, resulting in jam-packed order books and often the unfortunate side effect of overproduction using weak plant material. From the 1920s to the

The successful crossing of European *Aster amellus* with Himalayan *A. thomsonii* by Swiss nurseryman Carl Ludwig Frikart around 1918 was an important breakthrough in breeding asters.

Symphyotrichum novi-belgii 'Marie Ballard', a double-flowered beauty named for Ernest Ballard's wife, and one of his latest cultivars.

Symphyotrichum novi-belgii 'Gurney Slade', a fine cultivar from George Chiswell raised during the 1960s from Ernest Ballard's stock.

mid-1960s, a number of nurseries and breeders produced these new cultivars and made the most of the aster boom years. Through all of this, surprisingly little work was done in the plants' native country, perhaps for the same reason that we in the United Kingdom do not spend a lot of time trying to breed new cultivars of cardamine. The majority of commercial breeding was concentrated on New York asters, with more than one thousand cultivars raised during their heyday, before suffering a major decline in popularity at the end of the 1960s resulting in the loss of many cultivars. The breeding of New England asters also reached a peak between the 1920s and 1960s; however, unlike the New York asters they have just carried on at a steady rate, their popularity only really beginning to increase since the early twenty-first century. The breeding of European asters is much the same story but on an even smaller scale, although breeders on the European continent have been much more enterprising in this area as with a number of the Asiatic species, including *Aster trinervius*.

In the latter part of the twentieth century, the most influential breeders were to be found on the European continent, particularly in Germany. A major leap forward in colour was seen in 1969 with the introduction of the vibrant 'Andenken an Alma Pötschke' and again in 1981 when the pure white 'Herbstschnee' was raised by Heinz Klose, replacing a selection from the wild introduced by Amos Perry which had been lost during the chaos of the 1940s. The early 1990s saw the introduction of 'Purple Dome'—found growing wild beside a road in Pennsylvania—the plant that is now arguably the driving force behind a revival in the breeding of New England asters. The story of 'Purple Dome' reflects the story of many of the introductions from America insofar as they have been discovered wild, or are chance seedlings or sports appearing in gardens. Recently, however, a more concerted effort to raise new cultivars has begun, with the 'Woods' range of *Symphyotrichum novi-belgii* and a number of new *S. laeve* cultivars such as 'Bluebird' being noteworthy examples.

The early 1990s saw the introduction of 'Purple Dome'—found growing wild beside a road in Pennsylvania—the plant that is now arguably the driving force behind a revival in the breeding of New England asters. The story of 'Purple Dome' reflects the story of many of the introductions from America insofar as they have been discovered wild, or are chance seedlings or sports appearing in gardens.

Recent years have also seen renewed interest in the small-flowered asters. New introductions are playing an increasingly important role alongside old favourites such as 'Little Carlow'. Among these new introductions, some notable advancements in colour have been achieved including the compact *Symphyotrichum* 'Pixie Dark Eyes'.

Plant breeders in the Netherlands, Germany, and Israel have been raising large numbers of asters specifically for cut flower production. Originally, the crosses were made between garden varieties already in existence, but current breeding has produced

Until the 1950s, Mrs. R. B. Pole of Lye End Nursery continued to select new varieties of *Symphyotrichum novae-angliae*. Among the few still available today is 'Lye End Beauty'.

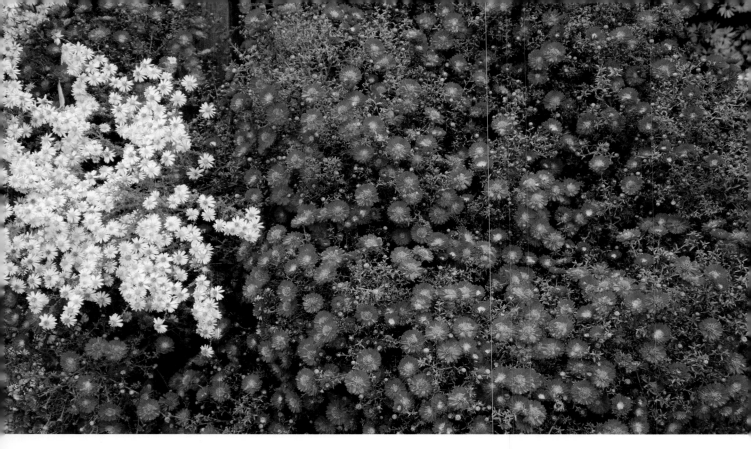

Symphyotrichum 'Pixie Dark Eyes', a dwarf hybrid of heath aster (*S. ericoides*) with red-purple flowers.

varieties much better suited to protected cultivation. Young plants are sent on licence to growers in many countries of the world so that huge quantities of asters (often known as "September Flowers") are sold through the flower markets and are available to florists year round. Commercially, this has been the most important development in aster breeding and sales since the early twentieth century. The value of this industry completely eclipses the sales of aster plants in traditional horticulture. Now, vases of asters can be seen in many houses that lack the gardens to grow the plants.

Although it has been a long slow road to asters becoming once again appreciated as valuable garden plants, renewed interest has been seen in the United Kingdom and the United States, including a glorious display at the Chicago Botanic Garden. And impressive new aster projects are also arising in less expected parts of the world, such as the gardens at Malmö Castle in Malmö, Sweden, where head gardener John Michael Taylor has created large double herbaceous borders packed with *Symphyotrichum novae-angliae*, *Aster amellus*, and *A. ×frikartii* cultivars. Most importantly, with so many species and cultivars to choose from, you are almost guaranteed to find an aster to suit your garden no matter where it is.

101
ASTERS
FOR THE
GARDEN

Small-flowered *Symphotrichum lateriflorum* 'Lady in Black' is very effective in front of autumn-colouring shrubs.

Many species and cultivars are available to grow in your garden. We chose the following selection because, between them, they display qualities to enhance so many different garden situations. Here is the place to find mounds of brightly coloured flowers, graceful pyramids of autumnal blooms, and a few tough customers to take care of difficult growing conditions. There are asters to grow in borders (formal and informal), cut flower borders, rock gardens, containers, and wildflower or meadow planting schemes.

Rather than trying to define the spread of each aster in this section, which is greatly influenced by growing conditions as well as inherited traits of a particular species or hybrid, we have generally described the spread by the nature of the clumps of shoots formed in one growing season. Most asters, including *Symphyotrichum novae-angliae*, tend to produce their shoots in compact clumps, which spread outwards at different rates according to variety. For instance, in ideal conditions *S. laeve* 'Vesta' would be described as a vigorous spreading clump, meaning it has the capability of producing a clump more than 60 cm (24 in.) across in one year. On the other end of the spectrum *S. lateriflorum* 'Lady in Black' produces a compact clump, meaning it will remain small, around 10 cm (4 in.) across.

Aster ageratoides 'Starshine'

Michaelmas daisy

This tough compact plant produces heads of crisp white flowers over a long period. The unusual toothed foliage creates mounds of mid-green in the early part of the year.

ZONES 4–8
HABIT AND SIZE Mounds of foliage are created by compact upright stems to 60 cm (24 in.) in height. The vigorous clumps are spreading but not invasive.
FLOWERS 2.5 cm (1 in.) across, from late summer until midautumn.
CULTIVATION Full sun or light shade, in any reasonable garden soil. Space 60 cm (24 in.) apart. Divide every four to five years. Mildew free.

ORIGIN Bred in the Netherlands.
LANDSCAPE AND DESIGN USES 'Starshine' provides an excellent display of pure white flowers over such a long time it makes it an ideal border plant combining well with other low-growing perennials. A particularly striking combination can be created by picking up the yellow of the disc with a pale yellow low-growing *Anthemis*.
SIMILAR PLANTS *Aster ageratoides* 'Asran' shares the same tough growth and toothed foliage as 'Starshine', but it produces pale lilac-blue flowers and grows taller, 75 cm (29 in.).

Aster alpinus
Alpine aster

The pretty lavender-blue, purple, or pink flowers are an especially welcome sight early in the season. Blooms are carried singularly on stems over spreading mats of foliage.

ZONES 3–8

HABIT AND SIZE Mats of foliage with the flowers on stems to 20–30 cm (8–12 in.) tall, forms compact clumps.

FLOWERS 5 cm (2 in.) across, from late spring into early summer.

CULTIVATION Full sun, in free draining soil. Space 60 cm (24 in.) apart. Divide every three to five years, or grow from seed (easier). Mildew free.

ORIGIN Throughout European mountains, West and Central Asia, Iran, Siberia, and western North America.

LANDSCAPE AND DESIGN USES Despite being short lived this aster is an invaluable addition to the late spring rock garden or raised bed, where it provides an essential burst of colour. It is also one of only three species of aster (the others are *A. amellus* and *Eurybia sibirica*) recommended as hardy enough to grow in Canada as a garden plant.

SIMILAR PLANTS *Aster alpinus* 'Dunkle Schöne' (synonym 'Dark Beauty') has deep purple flowers, 30 cm (12 in.) tall. *Aster* ×*alpellus*, a natural hybrid between *A. alpinus* and *A. amellus*, produces lavender-blue flowers and grows to 20–30 cm (8–12 in.) tall; it is very similar to *A. alpinus* but more reliably hardy.

Aster amellus 'Brilliant'

Italian starwort, European Michaelmas daisy

Of all the varieties of *A. amellus* on the market, 'Brilliant' is one of the most vigorous, a trait which is always desirable. The bright purple-pink, single flowers are produced abundantly from late summer until the frosts arrive. The foliage is green and quite hairy, but perhaps the most distinctive feature is the extremely upright habit.

ZONES 3–10
HABIT AND SIZE Upright sprays above strong but compact woody clumps to 60 cm (24 in.) tall.
FLOWERS 5.7 cm (2.25 in.) across, from late summer to midautumn.

CULTIVATION Full sun to light shade, in alkaline, well-drained, average soil. Space up to 60 cm (24 in.) apart. Divide every three to four years in spring. Resistant to powdery mildew.
ORIGIN Bred by Thomas Carlile in about 1953.
LANDSCAPE AND DESIGN USES 'Brilliant' works well in mixed borders where it can make substantial clumps providing a vibrant splash of colour late into the season. Like many of the *A. amellus* cultivars, the long flowering season of 'Brilliant' allows it to be combined with late summer herbaceous plants, such as *Anthemis*, as well as late autumn flowerers such as small-flowered *Symphyotrichum* cultivars.
SIMILAR PLANTS *Aster amellus* 'King George', which should be used as the standard for the species.

Aster amellus 'Forncett Flourish'

Italian starwort, European Michaelmas daisy

To read a basic description of this plant with its medium-sized, violet-blue single flowers you would not be greatly excited, thinking it much the same as many others. However, the vibrancy in colour of the flowers and their neat rounded shape make an astounding overall impression. The upright growth is reminiscent of 'Brilliant' but somehow the whole plant appears much tidier.

ZONES 3–10
HABIT AND SIZE Upright sprays from neat clumps, 60 cm (24 in.) in height and about 30 cm (12 in.) across during flowering.
FLOWERS 6.3 cm (2.5 in.) across, from early to midautumn.

CULTIVATION Full sun to light shade, in alkaline, well-drained, average soil. Space 30 cm (12 in.) apart. Divide every three to four years in spring. Resistant to powdery mildew.
ORIGIN A relatively modern cultivar bred by John Metcalfe at Four Seasons Nursery, United Kingdom, in 1996.
LANDSCAPE AND DESIGN USES A strong and floriferous constitution means that this cultivar makes a good specimen plant as well as being impressive as a group. Its relatively compact growth allows it to be used to the front of mixed borders where it is suited to both formal and naturalistic styles of planting.
SIMILAR PLANTS Aster amellus 'King George', which should be used as the standard for the species.

Aster amellus 'Gründer'

Italian starwort, European Michaelmas daisy

This is a distinctive statuesque variety of *A. amellus*. It produces large deep lavender-blue flowers with broad petals and a well-defined golden disc.

ZONES 3–10

HABIT AND SIZE Upright, strong bushy sprays from compact clumps, 80 cm (32 in.) in height and about 60 cm (24 in.) across during flowering.

FLOWERS 7.5 cm (3 in.) across, from early to midautumn.

CULTIVATION Full sun to light shade, in alkaline, well-drained, average soil. Space up to 45 cm (18 in.) apart. Divide every three to four years in spring. Resistant to powdery mildew.

ORIGIN From the name, it can be assumed that this cultivar is German in origin, and we know that it has been commercially available for a number of years. However, the exact details of the breeding and dates are unknown.

LANDSCAPE AND DESIGN USES Like many of the *A. amellus* cultivars 'Gründer' is suited to formal and informal situations and is much loved by insects. Try a combination using architectural plants to highlight the magnificent size of the luxurious blooms. Superb as a cut flower.

SIMILAR PLANTS *Aster amellus* 'King George', the standard for this species, is smaller in stature than 'Gründer' and has a more open flower form.

Aster amellus 'King George'

Italian starwort, European Michaelmas daisy

The size and richness of the purple-blue blooms make this a showier choice than many of the other *A. amellus* cultivars. The single flowers with their contrasting yellow discs are strikingly held on sturdy, well-branched sprays. The foliage is grey-green and distinctly hairy.

ZONES 3–10

HABIT AND SIZE Sprays produced from strong but compact woody clumps, 60 cm (24 in.) tall and 35 cm (14 in.) across during flowering.

FLOWERS 7 cm (2.75 in.) across, from late summer to midautumn.

CULTIVATION Full sun to light shade, in alkaline, well-drained, average soil. Space up to 45 cm (18 in.) apart. Divide every three to four years in spring. Resistant to powdery mildew.

ORIGIN Introduced from Germany by Amos Perry in 1914, where it had apparently been called 'Kaiser Wilhelm'. Due to the impending arrival of the First World War, it received a rapid name change.

LANDSCAPE AND DESIGN USES This plant is versatile enough to work in modern naturalistic plantings, the more traditional cottage garden designs, and even formal borders. A particularly striking arrangement can be formed when using this aster against formal hedging or topiary where the magnificent blooms are shown to perfection against the clean lines of the manicured shrubs. The large central discs of 'King George', as with most *A. amellus* cultivars, are perfect feeding stations for insects, such as the tortoiseshell butterfly seen in the photo.

SIMILAR PLANTS Both *Aster amellus* 'Sternkugel' with its large light violet-blue flowers, 50 cm (20 in.) tall, and *A. amellus* 'Vanity', which has large purple-blue flowers on slightly lax sprays, 45 cm (18 in.) tall, are very similar to 'King George' with just the slight colour and height varations.

Aster amellus 'Nocturne'

Italian starwort, European Michaelmas daisy

The large lilac flowers of 'Nocturne' have quite a distinctive colour among the A. *amellus* cultivars. Flowers are held on very strong sprays.

ZONES 3–10
HABIT AND SIZE Upright, strong, well-branched sprays from compact clumps, 80 cm (32 in.) in height and about 50 cm (20 in.) across during flowering.
FLOWERS 7 cm (2.75 in.) across, from early to midautumn.
CULTIVATION Full sun to light shade, in alkaline, well-drained, average soil. Space up to 45 cm (18 in.) apart. Divide every three to four years in spring. Resistant to powdery mildew.
ORIGIN Introduced by Alan Bloom around 1955, at his nurseries in Bressingham, United Kingdom.
LANDSCAPE AND DESIGN USES This selection is well suited to use in formal borders and cottage garden-style plantings. The relatively long straight stems make it ideal as a cut flower.
SIMILAR PLANTS *Aster amellus* 'Gründer' has similarly large flowers but they are deep lavender-blue; it grows 80 cm (32 in.) tall.

Aster amellus 'Rosa Erfüllung'

Italian starwort,
European Michaelmas daisy
SYNONYM *Aster amellus* 'Pink Zenith'

A profusion of beautiful bright purple-pink flowers on compact growth put on an excellent show from early autumn.

ZONES 3–10
HABIT AND SIZE Erect, bushy sprays from vigorous clumps, 50 cm (20 in.) in height and about 50 cm (20 in.) across during flowering.
FLOWERS 5 cm (2 in.) across, from early to midautumn.
CULTIVATION Full sun to light shade, in alkaline, well-drained, average soil. Space up to 45 cm (18 in.) apart. Divide every three to four years in spring. Resistant to powdery mildew.
ORIGIN Bred by Karl Foerster in Germany.
LANDSCAPE AND DESIGN USES This cultivar is suitable to use as a specimen plant or to form a purple-pink drift in early autumn. For a striking display, combine with silver-foliaged plants (such as *Stachys byzantina* 'Silver Carpet' or *Artemisia stelleriana* 'Boughton Silver') or set against dark foliage such as *Sedum telephium* subsp. *maximum* 'Atropurpureum'. Just make sure the companion growth is not so exuberant as to swamp the asters.
SIMILAR PLANTS Several *A. amellus* cultivars are similar to 'Rosa Erfüllung'. 'Jacqueline Genebrier' has rich purple-pink flowers, 75 cm (30 in.) tall. 'Peach Blossom' has weak sprays with pale purple-pink blooms, 40 cm (16 in.) tall. 'Mrs Ralph Woods' is a bright purple-pink, 60 cm (24 in.) tall.

Aster amellus 'Rudolf Goethe'

Italian starwort, European Michaelmas daisy

This cultivar stands out by exhibiting qualities that are more readily associated with hybrids such as *A. ×frikartii*. 'Rudolf Goethe' makes vigorous healthy clumps, from which broad pale green leaves and well-branched sprays of pale lavender are borne from late summer. The flowers themselves have well-spaced petals and in many ways closely resemble *A. ×frikartii* 'Mönch', making it a more compact alternative.

ZONES 3–10

HABIT AND SIZE Spreading sprays from vigorous clumps, 75 cm (30 in.) in height and about 60 cm (24 in.) across during flowering.

FLOWERS 6.3 cm (2.5 in.) across, from late summer to midautumn.

CULTIVATION Full sun to light shade, in alkaline, well-drained, average soil. Space 60 cm (24 in.) apart. Divide every three to four years in spring. Resistant to powdery mildew.

ORIGIN Bred by Georg Arends in 1914.

LANDSCAPE AND DESIGN USES Tougher than the average *A. amellus*, 'Rudolf Goethe' is easier to use in mixed borders, particularly of a less formal nature, where it is more likely to hold its own among tough companion plants.

SIMILAR PLANTS *Aster ×frikartii* 'Mönch' produces similar lavender-blue flowers over a longer period but is quite a bit taller at 90 cm (36 in.) tall.

Aster amellus 'Veilchenkönigin'

Italian starwort, European Michaelmas daisy
SYNONYM *Aster amellus* 'Violet Queen'

Being compact both in habit and flowering size in no way detracts from the effective show that is presented by the deep violet flowers of this plant. These richly coloured flowers are held on erect bushy sprays, and since the leaves are also deep green, the entire plant appears to glow. The flower rays usually show a tendency to form a whorl to the right.

ZONES 3–10
HABIT AND SIZE Well-branched sprays from small clumps, 40 cm (16 in.) in height and just under 30 cm (12 in.) across during flowering.
FLOWERS 5 cm (2 in.) across, from early to midautumn.
CULTIVATION Full sun to light shade, in alkaline, well-drained, average soil. Space up to 45 cm (18 in.) apart. Divide every three to four years in spring. Resistant to powdery mildew.

ORIGIN Bred in Germany by Karl Foerster in 1956.
LANDSCAPE AND DESIGN USES The name Karl Foerster is synonymous with the German movement of naturalistic planting, which today is so familiar to us. Foerster was famed for breeding plants that worked well when planted en masse, particularly grasses. In much the same way, this aster looks stunning planted in large groups and the starry flowers are perfectly offset by the architectural forms of grasses such as *Stipa tenuissima*. However, 'Veilchenkönigin' does not appreciate being crowded so avoid planting vigorous growers nearby.
SIMILAR PLANTS Two other *A. amellus* cultivars produce small violet flowers: 'Weltfriede' grows to 50 cm (20 in.) tall, and 'Kobold' to 40 cm (16 in.) tall. 'Kobold' is rather weak.

Aster 'Cotswold Gem'

Michaelmas daisy

Good pale purple-pink flowers on strong upright growth borne over an exceptionally long period. The colour becomes more intense when grown in partial shade.

ZONES 6–10

HABIT AND SIZE Sturdy sprays from strong compact clumps, 50–60 cm (20–24 in.) tall.

FLOWERS 4 cm (1.5 in.) across, from late summer to midautumn.

CULTIVATION Full sun or part-shade, in alkaline, well-drained, average soil. Space up to 45 cm (18 in.) apart. Divide every three to four years in spring. Resistant to powdery mildew.

ORIGIN Bred by Bob Brown of Cotswold Garden Flowers in the United Kingdom this lovely cultivar is a hybrid between *A. pyrenaeus* 'Lutetia' and most likely an *A. amellus*.

LANDSCAPE AND DESIGN USES Because of the extra vigour this hybrid has when compared to *A. amellus* cultivars it is possible to use in front of shrubs such as purple-leaved *Berberis* or the blue autumn-flowering *Ceanothus* ×*delilianus* 'Gloire de Versailles'.

SIMILAR PLANTS *Aster amellus* 'Rosa Erfüllung' is bright purple-pink, 50 cm (20 in.) tall, and *A. pyrenaeus* 'Lutetia' is an even paler lilac-blue, 60 cm (24 in.) in height, with a much laxer habit.

Aster ×frikartii 'Flora's Delight'

Michaelmas daisy

The earliest flowering variety of what can be called a Michaelmas daisy, 'Flora's Delight' starts to produce compact lilac flowers as soon as early summer. As the flowers age they tend to fade, but each individual is quickly replaced, and with deadheading the plant can look fantastic right into midautumn.

ZONES 6–10

HABIT AND SIZE Relatively weak clumps produce a number of well-branched sprays, 45 cm (18 in.) high.

FLOWERS 4 cm (1.5 in.) across, from early summer into midautumn.

CULTIVATION Full sun, in alkaline, well-drained, average soil. Space up to 30 cm (12 in.) apart. Divide every three to four years in spring. Resistant to powdery mildew.

ORIGIN A hybrid between A. amellus 'Sonia' and A. thomsonii 'Nanus', raised by renowned plantsman Alan Bloom in 1963.

LANDSCAPE AND DESIGN USES Unlike the other A. ×frikartii hybrids this plant is not very strong and will not thrive if it is overcrowded or allowed to become too wet in winter. Pot culture is in fact the preferable way to cultivate it, the disease resistance and compact nature of the plant making it perfect as a summer to autumn patio display. However, it will also thrive in raised beds and looks perfect nestled between rocks in the lower levels of rock gardens.

SIMILAR PLANTS Aster thomsonii produces lavender-blue flowers from late summer, 60 cm (24 in.) tall.

Aster ×frikartii 'Jungfrau'
Michaelmas daisy

Large violet-blue flowers are borne in profusion on strong, upright sprays. This cultivar is comparatively neglected among the A. ×frikartii hybrids yet the plant has a compact stature that means staking does not have to be considered. The flowers are significantly deeper in colour than 'Mönch' or 'Wunder von Stäfa' and much neater, presenting an attractive rounded flower to the world.

ZONES 4–10

HABIT AND SIZE Strong compact clumps produce numerous upright sprays, 60–70 cm (24–28 in.) high by 38 cm (15 in.) across during flowering.

FLOWERS 5 cm (2 in.) across, from late summer to midautumn.

CULTIVATION Full sun to part day shade, in alkaline, well-drained, average soil. Space 30 cm (12 in.) apart. Divide in spring, when about 2.5 cm (1 in.) of growth is showing, every three to five years. Resistant to powdery mildew.

ORIGIN This was one of the first batches of hybrids produced by Swiss nurseryman Frikart, in 1918. The other two were 'Mönch' and 'Eiger', all named for the peaks visible from his nursery.

LANDSCAPE AND DESIGN USES Like all the A. ×frikartii cultivars, this works well in a mixed border or in front of shrubs. One of my father's favourite combinations is with Coreopsis verticillata 'Moonbeam'. Another striking combination would be using 'Jungfrau' with Anemone ×hybrida 'Andrea Atkinson', Miscanthus sinensis 'Rotsilber', and Crocosmia ×crocosmiiflora 'Coleton Fishacre'. The shapes and colours are complementary and provide colour and interest for months.

SIMILAR PLANTS Aster ×frikartii 'Eiger' is also violet-blue, but flowers are not produced as freely as 'Jungfrau' and it is a much weaker plant, 80 cm (32 in.) tall.

Aster ×*frikartii* 'Wunder von Stäfa'

Michaelmas daisy

Large lavender-blue flowers are produced in quantity on strong, upright, well-branched sprays from midsummer until the first of the heavy frosts. This cultivar is usually over-looked by the media in preference for its older sibling *A.* ×*fri-kartii* 'Mönch'. Both look very similar and in all truth are frequently muddled in the trade; whichever one you end up with will be a superb garden plant. However, as a personal preference I (Helen) would tend towards 'Wunder von Stäfa' because of its slightly more compact and self-supporting habit.

ZONES 4–10

HABIT AND SIZE Strong compact clumps produce numerous well-branched sprays, 85 cm (33 in.) high by 76 cm (30 in.) across during flowering.

FLOWERS 8 cm (3 in.) across with 50 relatively broad petals, borne from midsummer to midautumn.

CULTIVATION Full sun, in alkaline, well-drained, average soil. Space 45 cm (18 in.) apart, but 80 cm (32 in.) to appreciate the individual shape. Divide every three to four years in spring. Resistant to powdery mildew.

ORIGIN One of the few successful hybridization efforts between *A. amellus* and *A. thomsonii*, achieved in 1924 by Swiss nurseryman Frikart.

LANDSCAPE AND DESIGN USES A superb specimen plant, when given the space it will produce a fantastic dome of flowers. The requirement for room around the plant to achieve this display means that they lend themselves to being mixed with early spring bulbs, where the very tight clumps will be virtually invisible. This is also of benefit to the bulbs as the late start of growth gives them a chance to gather all the light needed before dormancy. As with all *A.* ×*frikartii* cultivars, 'Wunder von Stäfa' also looks fantastic when planted in large groups.

SIMILAR PLANTS *Aster* ×*frikartii* 'Mönch' has lavender-blue flowers produced over a long season and looks virtually identical to 'Wunder von Stäfa'. The main difference is that 'Mönch' has a more upright habit, to 90 cm (36 in.) tall, meaning it can require some support.

Aster peduncularis

Michaelmas daisy
SYNONYM *Aster asperulus*

This unique Himalayan aster has distinctive broad, heart-shaped, hairy foliage above which large single lavender-blue flowers are carried individually on strong dark stems.

ZONES 4–8

HABIT AND SIZE Mounds are formed by the broad foliage to 60 cm (24 in.) in height; the flowers are then carried above this on stems up to 10 cm (4 in.) long. The clump will slowly spread via thick shoots reaching a width of approximately 40 cm (16 in.) in two years.

FLOWERS 5 cm (2 in.) across, usually starting in midsummer and continuing into midautumn.

CULTIVATION Full sun or light shade, in any reasonable garden soil. Space 60 cm (24 in.) apart. Divide every three to five years. Mildew free.

ORIGIN A species found wild in the Himalayas, now becoming increasingly popular as a garden plant.

LANDSCAPE AND DESIGN USES The long flowering season and relatively compact nature of this plant make it an ideal garden plant. The dense form of A. *peduncularis* means it works well with more structural plants such as *Persicaria amplexicaulis* 'Firetail' or any of the shorter *Panicum* cultivars available.

SIMILAR PLANTS The flowers closely resemble those of A. ×*frikartii* 'Wunder von Stäfa', as does the long flowering season, but the overall appearance of the plant is quite different.

Aster pyrenaeus 'Lutetia'

Michaelmas daisy

The palest lilac-blue starry flowers are freely produced on well-branched, arching sprays.

ZONES 6–10

HABIT AND SIZE Sturdy sprays with spreading branches from vigorous but compact clumps, 60 cm (24 in.) by up to 102 cm (40 in.).

FLOWERS 4 cm (1.5 in.) across, from late summer to midautumn.

CULTIVATION Full sun, in alkaline, well-drained, average soil. Space at least 60 cm (24 in.) apart. Divide every three to four years in spring. Resistant to powdery mildew.

ORIGIN This cultivar was bred in 1912 by the French nursery company Cayeaux (now famous for irises) and has been catalogued as both *A. amellus* and *A. ×frikartii*. It certainly exhibits traits of hybrid vigour which make the suggestions of a cross with *A. thomsonii* seem likely.

LANDSCAPE AND DESIGN USES This plant works well as a specimen plant because the sprays are lax enough that it will spill over the edge of raised beds, but at the same time, it makes a superb front-of-border plant. With its pale colour, it works as a foil for bolder coloured plants such as *Nerine bowdenii*, which is quite happy growing through it. Equally, it is neutral enough to work in subtle planting schemes with certain finer grasses or *Perovskia atriplicifolia* 'Blue Spire'. Maintenance is virtually non-existent since staking is a pointless and destructive exercise with this plant.

SIMILAR PLANTS *Aster* 'Cotswold Gem' is purple-pink, 50–60 cm (20–24 in.) in height with a more upright habit than 'Lutetia'.

Aster sedifolius

Michaelmas daisy
SYNONYMS *Aster acris, Galatella sedifolia*

Although each individual flower is rather small with very few petals, the overall display in early autumn is stunning. The flowers have eight to twelve petals, which are a warm lavender-blue colour with small but prominent pale yellow discs; these are massed on upright branches above fine ferny foliage. During warm weather, a distinct honey scent arises from this aster and it is much loved by insects. It should be mentioned that paler shades are often found in the wild, the warm lavender-blue has been selected for the garden-grown variety.

ZONES 6–8
HABIT AND SIZE Upright growth from strong compact clumps, occasionally reaching 90 cm (36 in.) in height, but often shorter.
FLOWERS 2.5 cm (1 in.) across, usually starting in late summer and continuing into early autumn.

CULTIVATION Full sun or light shade, in any reasonable well-drained garden soil. Space 50 cm (20 in.) apart. Divide every four to five years. Mildew free.
ORIGIN Introduced into the United Kingdom from Europe and northern Asia more than 250 years ago.
LANDSCAPE AND DESIGN USES So long as this plant is provided with a well-drained position it will thrive for many years gradually making a larger clump without the need for division. Given the right soil it is a superb border plant, however, the weight of flowers at the tops of the stems necessitates discreet staking or the use of sturdy planting companions such as *Sedum* 'Mr Goodbud' to support them.
SIMILAR PLANTS *Aster sedifolius* 'Nanus' is a more compact version of *A. sedifolius* reaching a maximum of 60 cm (24 in.) in height.

Aster tongolensis 'Napsbury'

East Indies aster
SYNONYM *Aster tongolensis* 'East India Aster'

This low-growing floriferous plant produces well-sized single flowers with distinctive large deep orange discs surrounded by violet-blue petals.

ZONES 6–8

HABIT AND SIZE Low-growing foliage forms compact spreading clumps from which numerous flowers are carried on stems to 30 cm (12 in.) in height.

FLOWERS 5 cm (2 in.) across, emerging from late spring into early summer.

CULTIVATION Full sun, in free draining fertile soil. Space 60 cm (24 in.) apart. Divide every three to five years. Mildew free.

ORIGIN The species *A. tongolensis* is a native of stony alpine meadows in the Himalayas from Western China into Nepal. This particular cultivar was bred in the United Kingdom.

LANDSCAPE AND DESIGN USES Excellent for creating late spring displays in raised beds, containers, or rock gardens. The flowers are also good as cut flowers.

SIMILAR PLANTS *Aster tongolensis* 'Berggarten' is lavender-blue with bright orange discs, 40 cm (16 in.) tall, and a fuller flower.

Aster trinervius var. *harae*

Michaelmas daisy

SYNONYMS *Aster microcephalus* var. *harae*,
Aster trinervius subsp. *ageratoides* var. *harae*

Masses of small deep violet flowers are produced on strong red-green stems with distinctive toothed foliage. The two truly unique features of this plant are the lateness of the flowers—which never open before midautumn and are often still going strong by early winter if the weather has been mild—and its ability to survive difficult growing conditions in full sun or light shade without ill effect.

ZONES 4–8
HABIT AND SIZE Upright growth from spreading clumps to 120 cm (4 ft.) tall.
FLOWERS 2.5 cm (1 in.) across, usually starting in midautumn and continuing into early winter.
CULTIVATION Full sun or light shade, in any reasonable garden soil or poor soil. Space 60 cm (24 in.) apart. Divide every four to five years. Mildew free.
ORIGIN Japan

LANDSCAPE AND DESIGN USES Although extremely tough it does look best in an open position where it will make a large clump, providing colour when everything else in the garden has finished. Because of the late flowering season it makes an effective combination when used among grasses such as *Miscanthus sinensis* 'Malepartus' whose pale seedheads form a beautiful background for the deep flowers. Similarly, shrubs that provide good winter stem colour can be combined with this aster to start the winter interest off with a bang. A few plants grown undercover, in a cold tunnel for example, can provide plentiful flowers for picking at Christmas. When grown in a semi-shaded position, it will exhibit a deeper colour.

SIMILAR PLANTS *Aster ageratoides* 'Ashvi' has pale stems, pale green foliage, and white flowers, reaching the same height as *A. trinervius* var. *harae*, 120 cm (4 ft.) tall.

Doellingeria umbellata

Flat-topped white aster
SYNONYM *Aster umbellatus*

Masses of small creamy white flowers are held in umbel-like groups at the top of sturdy upright stems. Each stem bears numerous upward facing branches of flowers at the top creating this flat arrangement at the apex of the plant. Following the flowers are beautiful silvery seedheads, as attractive as the flowers themselves, which carry the interest on into the early winter.

ZONES 4–8
HABIT AND SIZE Upright sturdy stems to 150 cm (59 in.) in height, from vigorous spreading clumps.
FLOWERS 2 cm (0.75 in.) across, from late summer to early autumn.
CULTIVATION Full sun, in any reasonable garden soil. Space 60 cm (24 in.) apart. Divide every three to five years. Mildew free.

ORIGIN Found growing wild in moist soil on woodland edges in Newfoundland to Minnesota and South to Georgia and Kentucky.
LANDSCAPE AND DESIGN USES This plant lends itself to making impressive swathes, working well in both relatively formal borders and more naturalistic styles of planting. It does tend, however, towards being rather vigorous so use in formal borders with caution. A wonderful combination can be achieved by growing the late flowering monkshood *Aconitum carmichaelii* 'Arendsii' alongside this aster but it also works well with many of the other large late-season perennials such as *Helianthus* 'The Monarch' and grasses such as *Molinia caerulea* subsp. *arundinacea* 'Transparent'.
SIMILAR PLANTS *Aster glehnii* subsp. *glehnii* has flowers that are smaller and a cleaner white; in fact, the whole plant is slightly finer in form than the sturdy *D. umbellata*. Reaches 150 cm (59 in.) in height.

Eurybia divaricata 'Eastern Star'

White wood aster
SYNONYM *Aster divaricatus* 'Eastern Star'

Small star-like white daisies are massed on rather lax wiry sprays, creating a low cloud-like effect. The flowers have six to nine petals unevenly spaced around small yellow discs, looking very much the part of small stars. Deep brown drooping wiry stems help to highlight the delicate flowers. The deep green foliage is very broad and toothed, again with dark stems. The most notable feature of all is that it will not only tolerate a shaded position but thrives there.

ZONES 4–8

HABIT AND SIZE The clumps stay compact but produce numerous sprawling sprays to about 50 cm (20 in.) in length, most of this length lends width rather than height to the plant.

FLOWERS 2.5 cm (1 in.) across, usually from late summer into early autumn.

CULTIVATION Full sun or shade, in any reasonable garden soil. Space 60 cm (24 in.) apart. Divide every three to five years. Free of powdery mildew.

ORIGIN Selected by American plantsman Roger Raiche.

LANDSCAPE AND DESIGN USES Ideal in large drifts as ground cover especially in those shadier places, provided they are not too dry, where the white flowers give a real lift. They are much loved by insects and have a light honey scent in warm weather. They can make striking combinations with foliage plants particularly deep purples or purple-reds.

SIMILAR PLANTS *Eurybia divaricata* is very close to 'Eastern Star' but is slightly smaller and less floriferous. *Eurybia schreberi* has masses of white flowers, held in heads on upright dark willowy stems, and grows taller than 'Eastern Star' at 60–90 cm (24–36 in.).

Eurybia ×herveyi 'Twilight'

Hervey's aster, large-leaf wood aster
SYNONYMS *Aster ×herveyi, A. macrophyllus* 'Twilight', *A.* 'Twilight', *A. ×commixtus, Eurybia macrophyllus* 'Twilight'

During late summer and early autumn, numerous lavender-blue flowers create seas of colour even in those dry shadier spots. The leaves form low mats with the flowers standing well above them.

ZONES 6–8
HABIT AND SIZE The stems are upright and do not need supports reaching 90 cm (36 in.) in height. The shallow-rooting clumps spread vigorously.
FLOWERS 3 cm (1.25 in.) across, from late summer to early autumn.
CULTIVATION Full sun or light shade, in any reasonable garden soil. Space 60 cm (24 in.) apart. Divide every three to five years. Mildew free.
ORIGIN 'Twilight' has been cultivated in the United Kingdom for many years but appears to be identical to the wild hybrid described as *E. ×herveyi*.

LANDSCAPE AND DESIGN USES Ideal for shadier spots in the garden where it is happy to create a mass of colour in the late summer. This particular shade of lavender-blue and the plant's tolerance for poor soil allows it to be combined very nicely with the more refined yellow crocosmias such as *Crocosmia ×crocosmiiflora* 'Honey Bells'. A nice structural contrast can be achieved using some of the larger sedums such as *Sedum* 'Joyce Henderson', whose purple-tinted foliage and purple-pink flowers look striking alongside 'Twilight'.
SIMILAR PLANTS *Eurybia macrophylla*, a large-leaved aster with flat heads of off-white flowers, is also shade tolerant and grows 80 cm (32 in.) tall. *Eurybia radula* produces violet-blue flowers in summer and grows up to 40 cm (16 in.) tall.

Eurybia sibirica
Siberian aster
SYNONYM *Aster sibiricus*

The clumps of this aster spread to form mounds of toothed foliage, which are then covered in pale lilac-blue flowers.

ZONES 3–8
HABIT AND SIZE Compact spreading clumps form mounds to 20 cm (8 in.) tall.
FLOWERS 2.5 cm (1 in.) across, opening from midsummer to early autumn.
CULTIVATION Full sun, in free draining garden soil, not too rich. Space 50 cm (20 in.) apart. Divide every three to five years. Mildew free.
ORIGIN This hardy little plant is circumpolar, found throughout Siberia, Canada, and Alaska, even above the Arctic Circle.
LANDSCAPE AND DESIGN USES Being fond of poorer conditions with good drainage makes this aster a very suitable candidate for the rock garden where it will happily flourish bringing colour for the late summer. When grown in the front of ordinary borders without overhanging growth the plant would be taller, while still maintaining the mounding habit. An interesting experiment would be to use this plant in gravel gardens where it could look charming.

Symphyotrichum 'Anja's Choice'

Michaelmas daisy
SYNONYM *Aster* 'Anja's Choice'

In midautumn spires of pearly pink flowers, excellent for cutting, are carried over compact clumps. The overall appearance of the plant shows the close relationship to *S. ericoides*.

ZONES 4–8
HABIT AND SIZE Upright spires from strong but compact clumps. Reaching a height of 120 cm (4 ft.) when established.
FLOWERS 4 cm (1.5 in.) across, in midautumn.
CULTIVATION Full sun, in good garden soil, which does not become waterlogged over winter. Space 60 cm (24 in.) apart. Divide every three to five years. Mildew resistant.

ORIGIN 'Anja's Choice' is one of a number of hybrids raised by Piet Oudolf in the Netherlands. The hybrids are achieved by crossing cut flower cultivars with garden cultivars, to get the beautiful flowering sprays but also strong enough growth to survive and thrive in gardens.
LANDSCAPE AND DESIGN USES Perfect as part of a mixed border, either as a specimen plant or as part of a group, the light colouring of the flowers provides a lift among the other heavy autumnal hues. Attractive foliage and a floriferous nature make 'Anja's Choice' a good selection for pot culture.
SIMILAR PLANTS *Symphyotrichum* 'Ochtendgloren' has purple-pink flowers and is slightly shorter than 'Anja's Choice' at 102 cm (40 in.).

Symphyotrichum 'Coombe Fishacre'

Michaelmas daisy
SYNONYM *Aster* 'Coombe Fishacre'

This excellent cultivar is smothered in small light purple-pink flowers with prominent yellow discs that rapidly fade to reddish from midautumn. The growth is compact but very bushy.

ZONES 4–8
HABIT AND SIZE Numerously branched stems arise from compact clumps to 90 cm (36 in.) in height.
FLOWERS 2 cm (0.75 in.) across, from early to midautumn.
CULTIVATION Full sun, in good garden soil. Space 60 cm (24 in.) apart. Divide every three to five years. Mildew resistant.

ORIGIN We know that this cultivar was raised at some point around 1920, however, the exact origins are still a mystery. Most likely, it is the result of a cross between *S. ericoides* and *S. novi-belgii*, despite the *S. lateriflorum* traits it seems to exhibit.

LANDSCAPE AND DESIGN USES As a border plant 'Coombe Fishacre' makes a gorgeous drift of colour in midautumn and is usually covered in insects. This pale purple-pink drift can make a bold statement when used as a foil for some of the deep purples such as *S. novae-angliae* 'Helen Picton'. Another effective use is planted in a large container on a sunny patio or courtyard.

Symphyotrichum cordifolium 'Chieftain'

Lowrie's blue wood aster, Michaelmas daisy
SYNONYM *Aster cordifolius* 'Chieftain'

A statuesque plant with large heart-shaped leaves. The whole plant erupts into a cloud of hazy lavender-blue during midautumn. The tiny flowers with yellow discs aging to purple are massed into tall spires.

ZONES 4–8
HABIT AND SIZE Strong, compact clumps produce sturdy stems that can easily reach 180 cm (6 ft.) in height, forming upright spires.
FLOWERS up to 2 cm (0.75 in.) across, from midautumn.
CULTIVATION Full sun, but will tolerate some light shade, in good garden soil, which does not become waterlogged over winter. Space 60 cm (24 in.) apart. Divide every three to five years. Mildew free.

ORIGIN The origin of this cultivar is unknown, but most seedlings come true, which suggests this may be a wild collected plant.
LANDSCAPE AND DESIGN USES An excellent choice for the very back of the border, making a striking foil for late-flowering perennial sunflowers such as *Helianthus* 'Lemon Queen'. A more unusual combination can be formed when using 'Chieftain' in front of autumn-colouring shrubs.
SIMILAR PLANTS *Symphyotrichum cordifolium* 'White Chief' has huge spires of small white flowers and can also reach 180 cm (6 ft.) in height.

Symphyotrichum cordifolium 'Elegans'

Lowrie's blue wood aster, Michaelmas daisy
SYNONYM *Aster cordifolius* 'Elegans'

This pretty cultivar is reliably one of the last to be in flower. The flowers are white with just a hint of pale violet, and when viewed as dense sprays, appear to have a silvery sheen.

ZONES 4–8
HABIT AND SIZE Strong, compact clumps produce upright, sturdy, dense sprays to a height of 120 cm (4 ft.) when established.
FLOWERS 1.3 cm (0.5 in.) across, from midautumn.
CULTIVATION Full sun, but will tolerate some light shade, in good garden soil, which does not become waterlogged over winter. Space 60 cm (24 in.) apart. Divide every three to five years. Mildew resistant.
ORIGIN This cultivar was raised in the United Kingdom before 1920. In fact it is often thought to probably be the nineteenth-century cultivar 'Albulus' but under an updated name.
LANDSCAPE AND DESIGN USES Despite the height of this cultivar, 'Elegans' can be used towards the middle of the border since it is quite light and airy. It makes a super foil for other late-season perennials, for example to soften the sometimes-harsh yellow of *Helianthus* 'Gullick's Variety', or to add interest in front of shrubs as they begin to turn in autumn.
SIMILAR PLANTS Two other cultivars of *S. cordifolium* that are both very similar to 'Elegans' as well as to each other are 'Silver Spray', with pale lavender flowers from early autumn, and 'Sweet Lavender', which is more of a true lavender and flowers in midautumn. Both reach 120 cm (4 ft.) in height like 'Elegans'.

Symphyotrichum ericoides 'Golden Spray'

White heath aster, Michaelmas daisy
SYNONYM *Aster ericoides* 'Golden Spray'

Small pure white flowers with a prominent golden disc smother the bushy sprays.

ZONES 4–8

HABIT AND SIZE Wiry, arching sprays are borne from moderately strong compact clumps. These sprays are densely packed and create beautiful mounds to a height of 90 cm (36 in.). The foliage is very fine.

FLOWERS 1.3 m (0.5 in.) across, from mid to late autumn.

CULTIVATION Full sun, in good moisture-retentive garden soil. Space 60 cm (24 in.) apart. Divide every three to five years. Resistant to powdery mildew, but may be troubled during very hard years.

ORIGIN Unknown

LANDSCAPE AND DESIGN USES This is an excellent all-round cultivar. The bushy mound form lends itself to both formal borders and the more naturalistic approach particularly in front of shrubs. It also looks superb in containers, and like so many small-flowered varieties makes an excellent cut flower.

SIMILAR PLANTS Other cultivars of *S. ericoides* include 'Blue Star', 'Brimstone', and 'Cinderella'. 'Blue Star' has small, rounded, lavender-blue flowers with growth very similar to 'Golden Spray'. 'Brimstone' has white flowers that emerge very late from yellow-tinted buds, and erect growth to 120 cm (4 ft.) tall. 'Cinderella' is close in appearance to 'Golden Spray' but has slightly less prominent discs and is generally a smaller and weaker plant, 75 cm (30 in.) in height.

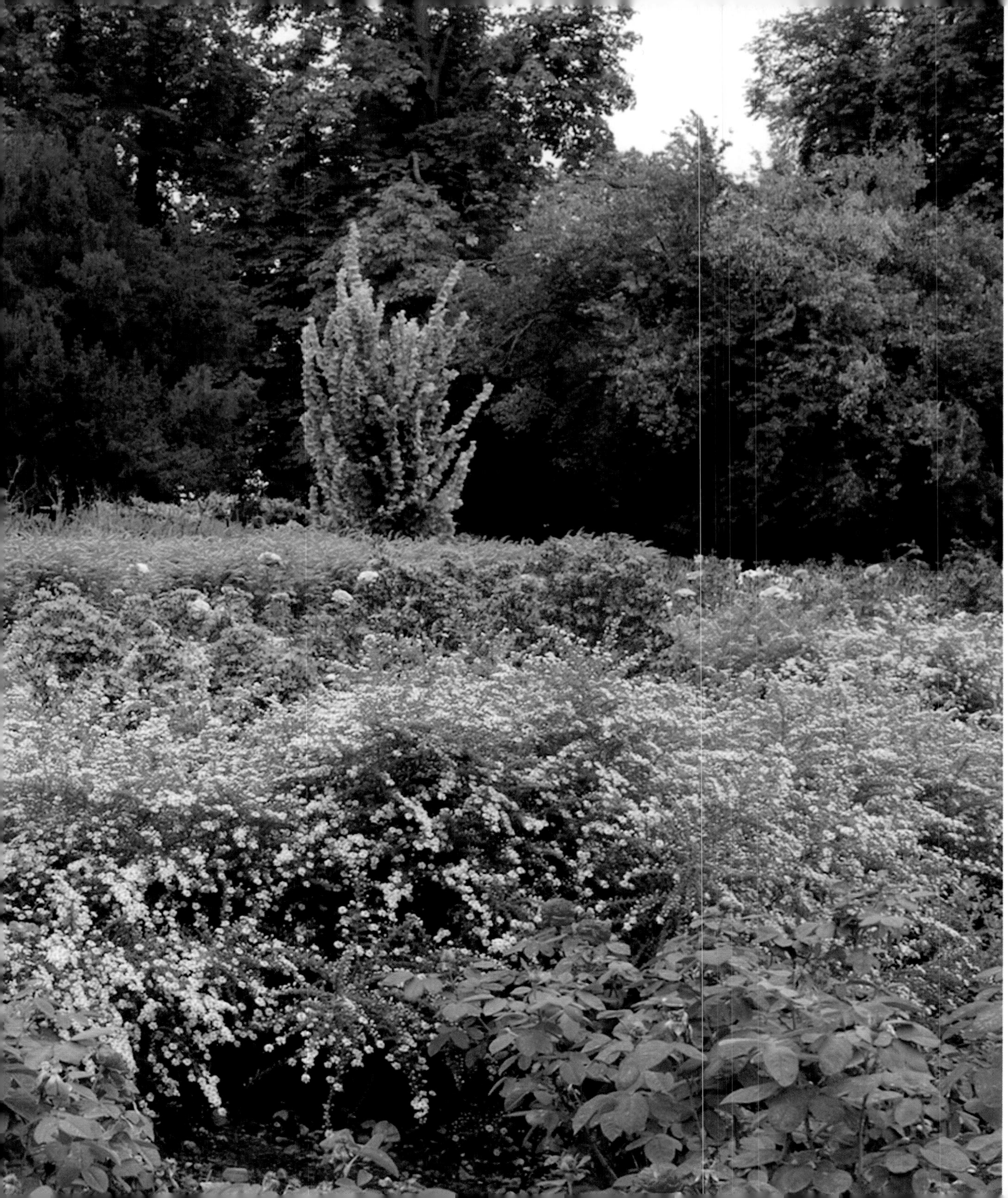

Symphyotrichum ericoides 'Pink Cloud'

White heath aster
SYNONYM *Aster ericoides*

To this day 'Pink Cloud' is still the best pink *S. ericoides* cultivar for garden use. The small pale purple-pink flowers smother the slightly arching bushy sprays. The dainty foliage has a distinct bronze tint in the spring, some of which is carried through the rest of the growing season.

ZONES 4–8
HABIT AND SIZE Forms more vigorous clumps than many other *S. ericoides* cultivars without being in danger of becoming invasive. Strong arching sprays reach a height of 90 cm (36 in.).
FLOWERS 1.3 cm (0.5 in.) across, from midautumn.
CULTIVATION Full sun, in good moisture-retentive garden soil. Space 60 cm (24 in.) apart. Divide every three to five years. Resistant to powdery mildew, but may be troubled during very hard years.
ORIGIN Unknown
LANDSCAPE AND DESIGN USES A plant with enough vigour to compete in the midst of a mixed border either as a specimen plant or in a swathe where it will create a pale pink cloud in late autumn. This plant is also tidy enough to use in containers or raised beds.

Symphyotrichum ericoides f. *prostratum* 'Snow Flurry'

Prostrate white heath aster
SYNONYM *Aster ericoides* f. *prostratus* 'Snow Flurry'

An unusual prostrate aster with fine ferny foliage. Instead of growing upright, the flowering stems grow flat to the ground forming a mat of mid-green throughout the summer. In mid-autumn, this mat of foliage becomes speckled with tiny white flowers, eventually turning the mat completely white due to the sheer quantity of flowers.

ZONES 4–8

HABIT AND SIZE The flowering stems can reach 30 cm (12 in.) in length and remain close to the ground, occasionally arching to 10 cm (4 in.) tall.

FLOWERS 1.3 cm (0.5 in.) across, from midautumn.

CULTIVATION Full sun, in good garden soil. Space 60 cm (24 in.) apart. Divide every three to five years. Mildew resistant.

ORIGIN The true origins of this plant remain a mystery. It was introduced to the United Kingdom from the United States by the well-known plantswoman Beth Chatto in 1983.

LANDSCAPE AND DESIGN USES This attractive plant lends itself to a range of uses, provided the flowering stems can grow over something other than bare soil where the moisture can damage them. In our own garden, we usually grow it on the edge of a dry stone wall where the stems spill down the front of the rock work. However, one of the most striking examples of how to use 'Snow Flurry' effectively is at Ragley Hall, Warwickshire, where it is used to edge the modernized rose garden. Here it is able to form an unobtrusive mat of foliage softening the square edge of the beds by spreading over onto the gravel path during the rose season and then add a wealth of interest in its own right when in flower during midautumn. The prostrate habit also makes it an ideal container plant.

Symphyotrichum 'Kylie'

New England aster
SYNONYMS *Aster* 'Kylie',
A. ×*amethystinus* 'Kylie',
A. novae-angliae 'Kylie'

Dainty pale purple-pink flowers on graceful wiry stems sprays make this aster stand out among its closest relatives, New England asters and small-flowered asters. The flowers are a deeper colour and slightly larger than you would normally find with *S. ericoides*, while they are much paler and smaller than *S. novae-angliae*.

ZONES 4–9
HABIT AND SIZE Compact woody clumps bear a multitude of wiry sprays, 120 cm (4 ft.) tall.
FLOWERS 1 cm (0.4 in.) across, from early autumn.
CULTIVATION Full sun, in any good garden soil. Space 60 cm (24 in.) apart. Divide every three to five years. Resistant to powdery mildew, occasionally rust can be a problem.
ORIGIN A hybrid between *S. novae-angliae* 'Andenken an Alma Pötschke' and *S. ericoides* 'White Heather' raised by Ronald Watts in 1990.
LANDSCAPE AND DESIGN USES An unusual cultivar for use in the border where it may need some support. Unlike most of the New England asters it also lasts very well when used as a cut flower.
SIMILAR PLANTS *Symphyotrichum ericoides* has tiny white flowers on wiry sprays, 102 cm (40 in.) tall.

Symphyotrichum laeve 'Calliope'

Smooth blue aster
SYNONYM *Aster laevis* 'Calliope'

This is probably the most iconic aster available, a robust and vigorous plant with deep purple willowy stems and purple-tinted foliage. Large lilac-purple flowers are freely produced in an open spire, lending this sizeable plant a surprisingly dainty air.

ZONES 4–8

HABIT AND SIZE The clumps are very strong and will spread vigorously in good conditions. Can easily reach a height of more than 2 m (6.5 ft.) in reasonable soil, less if the soil is poor.

FLOWERS 4.5 cm (1.75 in.) across, from midautumn

CULTIVATION Full sun, in any reasonable garden soil. Space 60 cm (24 in.) if dividing regularly, up to 90 cm (36 in.) apart for long-term naturalistic planting schemes. Divide every one to five years. Resistant to powdery mildew.

ORIGIN Raised before 1892 in the RHS Gardens at Chiswick, London.

LANDSCAPE AND DESIGN USES 'Calliope' lends itself to more wild or naturalistic planting schemes, where it can form large clumps and intermingle with other tough perennials or shrubs. The dark stems even without flowers look striking against a whole range of perennial grasses including a personal favourite of mine *Miscanthus sinensis* var. *condensatus* 'Cosmopolitan' with statuesque green and white variegated foliage.

SIMILAR PLANTS Three other cultivars of *S. laeve* include 'Bluebird', 'Nightshade', and 'Vesta'. 'Bluebird' has good lavender-blue flowers of a smaller size than 'Calliope', sprays are pyramidal in shape, 120 cm (4 ft.) tall. 'Nightshade' also produces lavender-blue flowers, well-branched sprays, and is 152 cm (5 ft.) tall. 'Vesta' has masses of pure white flowers, dark willowy stems, and very vigorous growth like 'Calliope'.

Symphyotrichum lateriflorum 'Lady in Black'

Calico aster, starved aster
SYNONYM *Aster lateriflorus* 'Lady in Black'

This graceful plant has beetroot-coloured spring foliage that is followed by deep purple stems and heavily tinted foliage. The lateral flowering branches are much more elegantly spaced than in the more diminutive cultivars such as 'Prince'. Late in the season, the purple stems are covered by huge numbers of small white flowers with deep purple-pink discs.

ZONES 4–8

HABIT AND SIZE From compact clumps the upright stems reach 130 cm (51 in.) in height.

FLOWERS 2 cm (0.75 in.) across, from midautumn.

CULTIVATION Full sun, in good garden soil. Space 60 cm (24 in.) apart. Divide every three to five years. Mildew resistant.

ORIGIN A newer addition to the range of asters available, 'Lady in Black' was raised by the Dutch plant breeders De Bloemenhoek in 1991.

LANDSCAPE AND DESIGN USES Superb as a border plant where it can add an architectural grace to even the smallest of areas whether formal or more naturalistic. The rich spring and early summer colour of the young foliage is excellent, making this plant a tougher and more slug proof alternative to *Lobelia cardinalis* 'Queen Victoria'. Like many of the small-flowered asters, 'Lady in Black' looks very effective in front of autumn-colouring shrubs and trees.

SIMILAR PLANTS *Symphyotrichum lateriflorum* is similar in habit to 'Lady in Black' but lacking the colouration of stem and foliage throughout the summer. However, this species will flower well even in light shade, 102 cm (40 in.) tall.

Symphyotrichum lateriflorum 'Prince'

Calico aster, starved aster
SYNONYM *Aster lateriflorus* 'Prince'

The feature that distinguishes this cultivar from other *S. lateriflorum* varieties is the intensity of the deep beetroot purple colouring to the spring growth, which remains over a long period of the growing season before eventually fading to a deep bronzed green. The flowering branches are held laterally from the main stem creating a very bushy plant. The small white flowers with prominent yellow discs, which rapidly fade to purple, are held on the top side of the branches, all clustering together to face upwards.

ZONES 4–8
HABIT AND SIZE The main stems grow from compact clumps to a height of 60 cm (24 in.). The spread of each plant is usually about 30 cm (12 in.) because of the horizontal habit of the flowering stems.
FLOWERS 1.3 cm (0.5 in.) across, from midautumn.

CULTIVATION Full sun, in good garden soil. Space 60 cm (24 in.) apart. Divide every three to five years. Mildew resistant.
ORIGIN Bred around 1970 by Eric Smith in the United Kingdom.
LANDSCAPE AND DESIGN USES Excellent both in the border and as a container plant, 'Prince' is certainly worthy of room in our overfilled gardens. A popular use for this cultivar is to edge a border or to create a faux hedge. Probably the best example of this latter use is at the famous garden Great Dixter, in East Sussex, although in this case it is actually 'Horizontalis' used. The same effect would be achieved using 'Prince'. In midautumn, the faux hedge is covered in honeybees and then it becomes a striking architectural feature when the first frosts arrive.
SIMILAR PLANTS *Symphyotrichum lateriflorum* 'Horizontalis' is very close to 'Prince' but has slightly paler foliage colour as the season advances.

Symphyotrichum 'Les Moutiers'

Michaelmas daisy
SYNONYM *Aster* 'Les Moutiers', *A.* 'Vasterival'

This plant is defined by its vigorous but non-invasive growth and a fine willowy nature. The pale pink flowers are borne in profusion on dark purple-red stems from early autumn, making a striking display. 'Les Moutiers', with its graceful sprays of pink stars, is finding its way into more and more English gardens each year.

ZONES 5–8
HABIT AND SIZE Strong willowy stems to at least 150 cm (59 in.) from spreading clumps.
FLOWERS 3.5 cm (1.5 in.) across, emerging from early autumn.
CULTIVATION Full sun, in good garden soil. Space 60 cm (24 in.) apart. Divide every three to five years. Mildew resistant.

ORIGIN 'Les Moutiers' came from a nursery in the northern French village of the same name.
LANDSCAPE AND DESIGN USES A super trouble-free border plant so long as it has to room to grow. 'Les Moutiers' is ideally suited to the less formal setting of the cottage garden style where it can form billowing masses of dancing pink stars. Also lovely in naturalistic plantings with perennial grasses.
SIMILAR PLANTS *Symphyotrichum laeve* 'Alba' also has dark slender stems to 120 cm (4 ft.), with white flowers in mid-autumn. *Symphyotrichum* 'Star of Chesters' is virtually identical to 'Les Moutiers'.

Symphyotrichum 'Little Carlow'

Michaelmas daisy
SYNONYM *Aster* 'Little Carlow'

An extensively planted aster in the United Kingdom, 'Little Carlow' has masses of bright lavender-blue flowers held in sumptuous panicles over broad deep green foliage.

ZONES 4–8

HABIT AND SIZE The clumps are strong and relatively compact, reaching 50 cm (20 in.) across in two years. These substantial clumps will produce numerous sprays giving a flowering plant diameter of 120 cm (4 ft.) across. The sturdy sprays are generally 120 cm (4 ft.) tall.

FLOWERS 2.5 cm (1 in.) across, from early autumn.

CULTIVATION Full sun, in good garden soil, which does not become waterlogged over winter. Divide every three to five years. Space at least 60 cm (24 in.) apart. Mildew resistant.

ORIGIN 'Little Carlow' is one of a select few surviving cultivars raised by Mrs. Thornely in Devizes, Wiltshire, during the 1930s and 1940s. The parentage of this hybrid is *S. cordifolium* and *S. novi-belgii*.

LANDSCAPE AND DESIGN USES Versatile is a word easily applied to 'Little Carlow'. It looks perfectly at home in a traditional herbaceous border, where the dense lavender-blue sprays will add that perfect dash of late-season colour, or in a less formal setting perhaps among grasses whose leaves are beginning to become burnished in the autumn light, creating a wonderful surround for the vivid flowers of this aster. As a specimen plant, it will make an impressive display, but with room, a swathe of 'Little Carlow' is unbeatable for length and quantity of flowering. Contrast the dense sprays of this aster with textures gleaned from other plants, such as the narrow foliage of kniphofias and the broad leaves of hostas such as 'Sum and Substance'. The latter also provides a glorious colour combination with the clear lavender-blue flowers of 'Little Carlow'.

Symphyotrichum novae-angliae 'Andenken an Alma Pötschke'

New England aster
SYNONYM *Aster novae-angliae* 'Andenken an Alma Pötschke'

The large strong sprays of vibrant cerise-pink, single flowers, one of the brightest colours found in asters, make this a striking plant. The appeal is helped by the relatively compact nature of the plant.

ZONES 4–9
HABIT AND SIZE Strong compact woody clumps produce numerous upright sprays, 90 cm (36 in.) tall.
FLOWERS 3.5 cm (1.5 in.) across, from early to midautumn.
CULTIVATION Full sun, in any good garden soil. Space 60 cm (24 in.) apart. Divide every two to five years. Resistant to powdery mildew.
ORIGIN Bred by Pötschke in Poland this cultivar was introduced to the United Kingdom in 1969 via Walters.

LANDSCAPE AND DESIGN USES The exceptional vibrancy can make this a difficult plant to place in the garden. However, that does not mean you should be scared of it. Backed by deep red-purple foliage, such as *Cotinus* 'Grace', or fronted by autumnal yellows, it can make a striking combination. Another option is to soften the impact by surrounding it with other lavender-blue or violet asters. Be bold and experiment with your own colour schemes.
SIMILAR PLANTS Other brightly coloured cultivars of *S. novae-angliae* include salmon-red 'Lachsglut', 152 cm (5 ft.) tall, and cerise 'Mary Young' with nicely shaped flowers, 120 cm (4 ft.) tall.

Symphyotrichum novae-angliae 'Barr's Pink'

New England aster

SYNONYM *Aster novae-angliae* 'Barr's Pink'

Uniquely large, deep lilac-pink flowers made even more striking by the presence of a distinctively prominent disc. The disc although initially yellow soon becomes a beautiful pale bronze making it a prime feature of the plant. In our garden, this always attracts the first crop of red admiral butterflies in the autumn.

ZONES 4–9
HABIT AND SIZE Vigorous woody clumps produce numerous strong, well-branched sprays, 152 cm (5 ft.) tall.
FLOWERS 5.5 cm (2.25 in.) across, from midautumn.
CULTIVATION Full sun, in any good garden soil. Space 60 cm (24 in.) or slightly more apart. Divide every two to five years. Resistant to powdery mildew.

ORIGIN Bred by Barrs of Taplow in the 1920s.
LANDSCAPE AND DESIGN USES 'Barr's Pink' is the very essence of a cottage garden plant and widely used in the United Kingdom. Suitable for numerous sunny positions throughout the garden with the ability to thrive even in harsh conditions. The height can be restrictive in smaller borders although a good clump provides colour and insect fodder late in the season as well as that extra bit of body so often missing in small areas.
SIMILAR PLANTS *Symphyotrichum novae-angliae* 'Pink Parfait' has large flowers in a paler shade of pink with slightly weaker growth to 120 cm (4 ft.) tall.

Symphyotrichum novae-angliae 'Harrington's Pink'

New England aster

SYNONYM *Aster novae-angliae* 'Harrington's Pink'

This tall and dignified aster is topped by delicate, pale, rose-pink flowers late in the season. The wiry stems tend towards needing some support unlike the majority of *S. novae-angliae* varieties, but the little extra effort is worth it for the unique colour.

ZONES 4–9

HABIT AND SIZE Moderately strong woody clumps produce wiry slightly lax stems and well-branched sprays, 152 cm (5 ft.) tall.

FLOWERS 2.5 cm (1 in.) across, from midautumn.

CULTIVATION Full sun, in any good garden soil. Space 60 cm (24 in.) apart. Divide every two to five years. Resistant to powdery mildew.

ORIGIN Discovered growing wild in Québec by farmer Millard Harrington, 'Harrington's Pink' was traded for a rare fern from the U.K. nursery Perry's of Enfield in 1943 and subsequently introduced for sale by that company.

LANDSCAPE AND DESIGN USES When used alongside the rich yellows of many late-season herbaceous plants, such as *Rudbeckia fulgida* var. *deamii*, a truly autumnal scene can be created. The gentle shade of pink melds with the shades of lavender, purple, and pink found among the other asters, so it is never difficult to find companion plants.

SIMILAR PLANTS Two cultivars of *S. novae-angliae* that are shorter than 'Harrington's Pink' but produce similar rose-pink flowers are 'Anabelle de Chazal', 102 cm (40 in.), and 'Rosa Sieger', 120 cm (4 ft.).

Symphyotrichum novae-angliae 'Helen Picton'

New England aster
SYNONYM *Aster novae-angliae* 'Helen Picton'

The intense violet-purple flowers are held in large sprays on top of woody upright stems. This cultivar's unique combination of colour depth and size in the flowers, along with a relatively compact nature, makes it stand out from others available on the market.

ZONES 4–9

HABIT AND SIZE Strong woody clumps produce numerous sturdy well-branched sprays, 100–120 cm (40–48 in.) tall.

FLOWERS 3.5 cm (1.5 in.) across, from early to midautumn.

CULTIVATION Full sun, in any good garden soil. Space 60 cm (24 in.) or slightly more apart. Divide every two to five years. Resistant to powdery mildew.

ORIGIN A first generation cross between *S. novae-angliae* 'Andenken an Alma Pötschke' and *S. novae-angliae* 'Purple Dome', raised by Paul Picton at Old Court Nurseries.

LANDSCAPE AND DESIGN USES Being woody at the base like all New England asters it is best used in a border, and not really suitable as a container plant. In the more traditional herbaceous border, the deep violet can be used effectively repeated down the length complementing the yellows that tend to dominate in the autumn. A brilliant example of this can be seen at Waterperry in Oxfordshire. In modern planting schemes, swathes of perennial grasses that are beginning to mellow into the straw-like shades of autumn can be given an easy boost of colour in midautumn with large clumps of this aster.

SIMILAR PLANTS *Symphyotrichum novae-angliae* 'Violetta' has deep violet-purple flowers (as expected), 150 cm (59 in.) tall, and *S. novae-angliae* 'Marina Wolkonsky' is deep violet, 120 cm (4 ft.) tall.

Symphyotrichum novae-angliae 'Herbstschnee'

New England aster

SYNONYM *Aster novae-angliae* 'Herbstschnee'

An entirely pale plant. The foliage is a very pale green unlike the normal foliage of New England asters. Above the foliage, white flowers with pale yellow centres are held on moderately strong sprays.

ZONES 4–9

HABIT AND SIZE Strong woody clumps produce sprays of compact branches, 120 cm (4 ft.) tall.

FLOWERS 4.5 cm (1.75 in.) across, in early to midautumn.

CULTIVATION Full sun, in any good garden soil. Space 60 cm (24 in.) apart. Divide every two to five years. Resistant to powdery mildew, occasionally rust can be a problem.

ORIGIN Originally from Germany, bred by H. Klose in 1981.

LANDSCAPE AND DESIGN USES White can always provide a useful break among the rich colours of other autumn herbaceous plants and the turning foliage of trees and shrubs. Being tall, this aster is ideal towards the back of borders particularly where the background is composed of deciduous trees and shrubs.

Symphyotrichum novae-angliae 'James Ritchie'
New England aster
SYNONYM *Aster novae-angliae* 'James Ritchie'

Relatively compact growth with good sprays of attractive bright purple-red single flowers. The compact but strong growth combined with the bold flower colour makes this plant an outstanding recent introduction.

ZONES 4–9
HABIT AND SIZE Woody clumps produce strong well-branched sprays, 100–120 cm (40–48 in.) tall.
FLOWERS 3.5 cm (1.5 in.) across, from early to midautumn.

CULTIVATION Full sun, in any good garden soil. Space 60 cm (24 in.) apart. Divide every two to five years. Resistant to powdery mildew.
ORIGIN Bred by Julie Ritchie of Hoo House Nurseries, Tewkesbury, United Kingdom, and named for her son.
LANDSCAPE AND DESIGN USES A useful addition to the mid-border where reds are often in short supply.
SIMILAR PLANTS *Symphyotrichum novae-angliae* 'Colwall Orbit' has vivid cerise rounded flowers with a prominent disc, 90 cm (36 in.) tall.

Symphyotrichum novae-angliae 'Lou Williams'

New England aster
SYNONYM *Aster novae-angliae* 'Lou Williams'

This plant is magnificent in its stature and self-supporting nature alone. The single purple-red flowers, which appear slightly paler due to the broad petals twisting to reveal the pallid reverse, are also large, adding to the overall effect.

ZONES 4–9
HABIT AND SIZE Compact woody clumps produce strong sprays, 180 cm (6 ft.) tall.
FLOWERS 5.5 cm (2.25 in.) across, from midautumn.
CULTIVATION Full sun, in any good garden soil. Space 60 cm (24 in.) apart. Divide every two to five years. Resistant to powdery mildew, occasionally rust can be a problem.
ORIGIN Bred in 1995 by John Williams in Warwickshire, United Kingdom.
LANDSCAPE AND DESIGN USES An excellent back-of-border plant that associates brilliantly with *Helianthus* particularly in the paler shades of lemon such as *H.* 'Lemon Queen' or the airy *Rudbeckia subtomentosa*. With the latter the reflexed petals and protruding cone contrast with the rather flat flowers of 'Lou Williams'. Away from the more formal borders, this aster will also stand its ground among the taller and more robust grasses in prairie-style plantings without becoming swamped.
SIMILAR PLANTS *Symphyotrichum novae-angliae* 'Lucida' is earlier to flower than 'Lou Williams' and has smaller purple-red flowers, 140 cm (55 in.) tall.

Symphyotrichum novae-angliae 'Mrs S. T. Wright'

New England aster
SYNONYM *Aster novae-angliae* 'Mrs S. T. Wright'

With sumptuous large lilac-blue flowers borne in profusion, this is one of the finest old varieties still in cultivation.

ZONES 4–9
HABIT AND SIZE Vigorous clumps produce strong sprays, 152 cm (5 ft.) tall.
FLOWERS 5 cm (2 in.) across, in early to midautumn.
CULTIVATION Full sun, in any good garden soil. Space 60 cm (24 in.) apart. Divide every two to five years. Resistant to powdery mildew.
ORIGIN Raised by H. J. Jones in the United Kingdom before 1907.
LANDSCAPE AND DESIGN USES Makes a substantial statement when used as a group in the border, and is always a favourite for attracting late butterflies.
SIMILAR PLANTS *Symphyotrichum novae-angliae* 'Mrs S. W. Stern', 120 cm (4 ft.), and *S. novae-angliae* 'Treasure', 140 cm (55 in.), both have large light violet flowers.

Symphyotrichum novae-angliae 'Primrose Upward'

New England aster
SYNONYM *Aster novae-angliae* 'Primrose Upward'

Although the bold purple-red flowers are similar in colour to a number of older varieties, this plant produces excellent reliable growth (good-sized flowers on strong self-supporting stems) and is highly recommended as a garden plant.

ZONES 4–9

HABIT AND SIZE Robust sprays from compact woody clumps, 120–140 cm (48–55 in.) tall.

FLOWERS 4.5 cm (1.75 in.) across, in early to midautumn.

CULTIVATION Full sun, in any good garden soil. Space 60 cm (24 in.) apart. Divide every two to five years. Resistant to powdery mildew.

ORIGIN A seedling from nurseryman Bob Brown, Evesham, United Kingdom. Thought most likely to be a cross between *S. novae-angliae* 'Sayer's Croft' and *S. novae-angliae* 'Septemberrubin'.

LANDSCAPE AND DESIGN USES A fine statement plant whether used singularly or as a group. Suited to use as a traditional back-of-border plant or in naturalistic plantings.

SIMILAR PLANTS Other rich purple-red cultivars of *S. novae-angliae* include 'Crimson Beauty', with flowers 2.5 cm (1 in.) across, 140 cm (55 in.) tall; 'Septemberrubin', 5 cm (2 in.) across, weak growth to 130 cm (51 in.); and 'Lucida', 2.5 cm (1 in.) across, strong upright growth to 140 cm (55 in.).

Symphyotrichum novae-angliae 'Purple Dome'

New England aster
SYNONYM *Aster novae-angliae* 'Purple Dome'

The outstanding feature of this plant is that it is so much shorter than most other New England asters. Large violet-purple flowers top these compact bushy domes of deep green foliage.

ZONES 4–9
HABIT AND SIZE Strong compact clumps produce numerous branching sprays, 60 cm (24 in.) tall.
FLOWERS 4 cm (1.5 in.) across, in midautumn.
CULTIVATION Full sun, in any good garden soil. Space 60 cm (24 in.) apart. Divide every two to five years. Resistant to powdery mildew, occasionally rust can be a problem.
ORIGIN This unique plant was found growing beside a road in the United States by R. Seip; Lighty and Simon introduced it to the United Kingdom around 1990.

LANDSCAPE AND DESIGN USES Dense dark green mounds of foliage throughout the summer do not lend themselves to being interesting, but when combined with something like the ethereal *Stipa tenuissima* as a backdrop, the shapes work well together and later the aster's violet-purple flowers are highlighted by the faded silvery seedheads of the grass. 'Purple Dome' can also be used as a container-grown specimen, which is unique among New England asters. When growing it this way, use a large container and plant something smaller around the edge to hide the bottom part of the stems.
SIMILAR PLANTS *Symphyotrichum novae-angliae* 'Vibrant Dome' has purple-pink flowers, 60 cm (24 in.) tall.

Symphyotrichum novae-angliae 'Rubinschatz'

New England aster
SYNONYM *Aster novae-angliae* 'Rubinschatz'

Fine deep purple-pink petals make up these unusual flowers, which have a startlingly rounded, flattened look when fully open. These are completed with a deep golden disc. The sprays are strong but because the individual branches are compact, the overall effect is very neat.

ZONES 4–9
HABIT AND SIZE Strong woody clumps bear numerous sprays, 130 cm (51 in.) tall.
FLOWERS 4.5 cm (1.75 in.) across, in early to midautumn.
CULTIVATION Full sun, in any good garden soil. Space at least 60 cm (24 in.) apart. Divide every two to five years. Resistant to powdery mildew, occasionally rust can be a problem.

ORIGIN Raised in Germany by the esteemed Karl Foerster in 1960.
LANDSCAPE AND DESIGN USES As with most New England asters 'Rubinschatz' is best suited to being grown in the middle or at the back of a border, combining well with other herbaceous plants, grasses, or even neat shrubs.
SIMILAR PLANTS Three cultivars of *S. novae-angliae* are worth noting here. 'Andenken an Paul Gerber' has deep purple-pink flowers, 140 cm (55 in.) tall. 'Quinton Menzies' and 'Lye End Beauty' both produce bright purple-pink flowers and grow to 140 cm (55 in.) tall.

Symphyotrichum novae-angliae 'St. Michael's'

New England aster
SYNONYM *Aster novae-angliae* 'St. Michael's'

The rich purple-blue flowers have significantly more petals than many New England asters, giving them a very luscious appearance. Such a full-bodied flower would perhaps lead one to suppose that it was not terribly weather resistant. However, the flowers remain open to face the elements even during the wettest autumn.

ZONES 4–9

HABIT AND SIZE Strong woody clumps bear a multitude of large upright sprays, 120 cm (4 ft.) tall.

FLOWERS 5 cm (2 in.) across, in early to midautumn.

CULTIVATION Full sun, in any good garden soil. Space 60 cm (24 in.) apart. Divide every two to five years. Resistant to powdery mildew.

ORIGIN Raised by Paul Picton in the United Kingdom and named for St. Michael's Hospice in Herefordshire.

LANDSCAPE AND DESIGN USES Perfectly suited to a sunny position in the middle or back of borders or as a mainstay of prairie-style plantings.

SIMILAR PLANTS *Symphyotrichum novae-angliae* 'Augusta' is pale purple-blue, 102 cm (40 in.) tall, and *S. novae-angliae* 'Barr's Blue' is purple-blue, 117 cm (46 in.) tall.

Symphyotrichum novi-belgii 'Anita Ballard'

New York aster, long leaf aster, traditional Michaelmas daisy
SYNONYM *Aster novi-belgii* 'Anita Ballard'

This is one of the earliest-flowering New York asters. The pale lavender-blue flowers have a good number of petals but are still considered single. The sprays are well branched and strong, forming a pyramidal shape, meaning that although tall, 'Anita Ballard' can be used further forward in the border if so desired.

ZONES 4–8
HABIT AND SIZE Vigorous clumps produce pyramidal sprays, 150–180 cm (5–6 ft.) tall.
FLOWERS 4.5 cm (1.75 in.) across, from early to midautumn.
CULTIVATION Full sun, in good moisture-retentive garden soil. Space 60 cm (24 in.) apart. Divide every one to three years. Susceptible to powdery mildew.
ORIGIN Bred by Ernest Ballard at Old Court Nurseries before 1935. This is viewed as part of the progression of Ballard's work being a significantly better colour than 'Beauty of Colwall', which was one of his earliest cultivars, but lacking the number of petals he later achieved.

LANDSCAPE AND DESIGN USES A grouping of this plant makes a bold statement in the border, the flowers contrasting beautifully with the architectural properties of the taller grasses such as *Miscanthus sinensis* 'Rotsilber'. Ernest Ballard wrote that the colour of this cultivar was an "exquisite wistful shade of soft china-blue," which is quite an apt if artistic description and helps the mind envisage how well it works among the warm colours of heleniums, particularly as a background for the bold red of *Helenium* 'Dunkle Pracht'. However, it also lends itself to a more mellow approach with the soft pink of the architectural *Althaea cannabina* or the cool citrus of *Helianthus* 'Lemon Queen'.
SIMILAR PLANTS *Symphyotrichum novi-belgii* 'Algar's Pride' produces single, lavender-blue flowers that are larger and appear later than those of 'Anita Ballard', 152 cm (5 ft.) tall. The flowers are often compared to those of *Aster* ×*frikartii* 'Mönch'.

Symphyotrichum novi-belgii 'Apollo'

New York aster, long leaf aster, traditional Michaelmas daisy
SYNONYM *Aster novi-belgii* 'Apollo'

This plant produces masses of white flowers, which cover the low doming growth, over a long period. Even when 'Apollo' appears to be in full flower, many buds can still be found beneath the first crop just waiting to emerge.

ZONES 4–8
HABIT AND SIZE Strong compact clumps produce numerous bushy sprays, 35 cm (14 in.) tall.
FLOWERS 2.5 cm (1 in.) across, in early to midautumn.
CULTIVATION Full sun, in good moisture-retentive garden soil. Space 45 cm (18 in.) apart. Divide every one to three years. Susceptible to powdery mildew, but rarely a problem.

ORIGIN Unknown
LANDSCAPE AND DESIGN USES As with all *S. novi-belgii* cultivars 'Apollo' looks most glorious when planted as a group. However, the doming form of the sprays means that it can also make a charming specimen plant for the front of small borders or as a container plant. In fact, a well-grown plant of 'Apollo' surrounded by annuals such as *Bacopa* 'Great Purple' makes a beautiful patio container full of colour from early summer until midautumn.
SIMILAR PLANTS *Symphyotrichum novi-belgii* 'Schneekissen' (synonym 'Snow Cushion') produces single white flowers, and *S. novi-belgii* 'Snow Sprite' has double white flowers. Both plants grow to 30 cm (12 in.) tall.

Symphyotrichum novi-belgii 'Autumn Beauty'

New York aster, long leaf aster, traditional Michaelmas daisy
SYNONYM *Aster novi-belgii* 'Autumn Beauty'

Living up to its name, 'Autumn Beauty' is hard to beat for length and lusciousness of flowering. The large flowers have a formal double shape and are a soft warm shade of lilac-blue, creating an almost tangible sense of gentle autumnal evenings.

ZONES 4–8

HABIT AND SIZE Vigorous clumps produce open sturdy sprays, 100 cm (39 in.) tall.

FLOWERS 5 cm (2 in.) across, from early to midautumn.

CULTIVATION Full sun, in good moisture-retentive garden soil. Space 50 cm (20 in.) apart. Divide every one to three years. Susceptible to powdery mildew.

ORIGIN Unknown

LANDSCAPE AND DESIGN USES Because this cultivar will flower right the way down the stem, it is one of the few taller varieties that can make a nice container-grown plant even if it is more than 90 cm (36 in.) tall. In the border, it puts on a solid show of colour over a significantly long period.

SIMILAR PLANTS *Symphyotrichum novi-belgii* 'Ada Ballard' produces double, lavender-blue flowers in early autumn, 100 cm (39 in.) tall.

Symphyotrichum novi-belgii 'Autumn Rose'

New York aster, long leaf aster, traditional Michaelmas daisy
SYNONYM *Aster novi-belgii* 'Autumn Rose'

This is among the latest-flowering *S. novi-belgii* varieties, invariably starting in midautumn and continuing until the heavy frosts. For us, the autumn of 2013 saw 'Autumn Rose' with flowers on in the first week of November. If taken individually, the flowers—mauve-pink, relatively small, and single—are not too impressive, but an entire clump is a welcome boost of colour very late in the season.

ZONES 4–8
HABIT AND SIZE Strong compact clumps produce upright but well-branched sprays, 120 cm (4 ft.) tall.
FLOWERS 4.5 cm (1.75 in.) across, in midautumn.
CULTIVATION Full sun, in good moisture-retentive garden soil. Space 50 cm (20 in.) apart. Divide every one to three years. Susceptible to powdery mildew.

ORIGIN Unknown
LANDSCAPE AND DESIGN USES The late-flowering nature of this cultivar is what makes it a must-have in the garden, but it is also what makes it harder to use. Ensuring that other late-flowering plants are nearby helps stop it looking out of place; consider using small-flowered *Symphyotrichum* varieties such as *S. lateriflorum* 'Lady in Black'. Another striking way to use it is against golden foliage, for instance *Cotinus coggygria* 'Ancot'. Alternatively, structural plants can also be used to fulfil the same purpose of highlighting the late-season colour without making it seem out of place.
SIMILAR PLANTS Also flowering in midautumn, *S. novi-belgii* 'Blue Gown' has deep lavender-blue flowers, 100 cm (39 in.).

Symphyotrichum novi-belgii 'Beechwood Charm'

New York aster, long leaf aster, traditional Michaelmas daisy
SYNONYM *Aster novi-belgii* 'Beechwood Charm'

Unlike the majority of New York asters where the individual flowers are substantial in size, 'Beechwood Charm' has distinctive, small, deep pink flowers. These flowers are produced prolifically with each spray being home to a large number of them. This serves to make the plant seem extremely floriferous.

ZONES 4–8
HABIT AND SIZE Stiff, well-branched, bushy sprays, 100 cm (39 in.) tall, from vigorous clumps.
FLOWERS 2.5 cm (1 in.) across, from early to midautumn.

CULTIVATION Full sun, in good moisture-retentive garden soil. Space 60 cm (24 in.) apart. Divide every one to three years. Susceptible to powdery mildew.
ORIGIN Many consider 'Beechwood Charm' as the best surviving cultivar bred by W. Wood. Raised around 1937.
LANDSCAPE AND DESIGN USES The floriferous nature of this aster makes it a superb cut flower as well as an excellent candidate for the border and for naturalistic planting schemes.
SIMILAR PLANTS *Symphyotrichum novi-belgii* 'Charles Wilson' has small, pale purple-red flowers from early autumn and grows as tall as 'Beechwood Charm'.

Symphyotrichum novi-belgii 'Blauglut'

New York aster, long leaf aster, traditional Michaelmas daisy
SYNONYMS *Aster novi-belgii* 'Blauglut'

Stormy violet flowers with discs that are small compared to the length of petals are produced freely on very bushy growth over a long season. Flowering sprays are formed right down the length of the stems and, given room, flowers will bloom from almost ground level to the top of the plant.

ZONES 4–8
HABIT AND SIZE Vigorous clumps produce bushy sprays, 90 cm (36 in.) tall.
FLOWERS 4.5 cm (1.75 in.) across, from early to midautumn.
CULTIVATION Full sun, in good moisture-retentive garden soil. Space 60 cm (24 in.) apart. Divide every one to three years. Sometimes susceptible to powdery mildew.
ORIGIN As the name suggests, 'Blauglut' is of German origin, bred by H. Klose in 1980.
LANDSCAPE AND DESIGN USES With flowers being produced from almost ground level upwards this aster can be used at the front of the border despite being quite tall, creating a striking cloud of flowers which looks super in front of pale pink or white shades. A favourite combination is to plant it in front of *S. novi-belgii* 'Lassie', which is slightly taller, creating a glorious riot of colour. Work this combination in with orange, red, or purple tinted foliage and you have an almost perfect autumn palette.
SIMILAR PLANTS *Symphyotrichum novi-belgii* 'Purple Dome' produces deep heather-purple flowers, and *S. novi-belgii* 'Pride of Colwall' has small, double, heather-purple flowers. Both flower from early autumn and grow to 90 cm (36 in.) tall.

Symphyotrichum novi-belgii 'Blue Lagoon'

New York aster, long leaf aster
SYNONYM *Aster novi-belgii* 'Blue Lagoon'

For a dwarf cultivar, this plant has exceptionally large flow-ers of an unusual shade of violet-blue. The foliage is also a distinctive deep green colour, as well as being broad for the overall size of plant, which emphasizes a very healthy look.

ZONES 4–8
HABIT AND SIZE Compact clumps produce upright sprays, 45 cm (18 in.) tall.
FLOWERS 5 cm (2 in.) across, from early to midautumn.
CULTIVATION Full sun, in good moisture-retentive garden soil. Space 40 cm (16 in.) apart. Divide every one to four years. Susceptible to powdery mildew, but not particu-larly troubled.

ORIGIN Unknown
LANDSCAPE AND DESIGN USES Able to thrive even if left undivided for a number of years, this aster is a suitable candidate for lower maintenance planting schemes, as well as ideal for traditional herbaceous borders. As with most New York asters, it performs best when planted in a group, where it will make a bold statement.
SIMILAR PLANTS *Symphyotrichum novi-belgii* 'Dora Chiswell' also has violet-blue flowers, 45 cm (18 in.) tall.

Symphyotrichum novi-belgii 'Chequers'

New York aster, long leaf aster
SYNONYM *Aster novi-belgii* 'Chequers'

This plant produces a richly coloured mass of deep violet-purple flowers with distinctive bright golden discs. Contrast is really the key to its effectiveness as the yellow and purple combination makes the entire plant glow. 'Chequers' is also a useful height: neither tall enough to be relegated to the back of the border nor so short that the front of the border is the only option.

ZONES 4–8
HABIT AND SIZE Bushy sprays from small clumps, 60 cm (24 in.) tall.
FLOWERS 4.5 cm (1.75 in.) across, from early to midautumn.
CULTIVATION Full sun, in good moisture-retentive garden soil. Space 40 cm (16 in.) apart. Divide every one to three years. Susceptible to powdery mildew.
ORIGIN Raised by R. Lidsey, 'Chequers' was introduced through Gayborder Nurseries around 1953.
LANDSCAPE AND DESIGN USES Most effective as a group, the larger the better. If you have the space, a small drift of this cultivar would look stunning among *Stipa tenuissima*. In a more formal setting, a smaller group works well when contrasted against white or pale pink flowers. Another idea for an excellent formal planting would be to use 'Chequers' as part of a modern parterre.
SIMILAR PLANTS *Symphyotrichum novi-belgii* 'Purple Dome' has deep heather-purple flowers, 90 cm (36 in.) tall.

Symphyotrichum novi-belgii 'Climax'

New York aster, long leaf aster, traditional Michaelmas daisy
SYNONYMS *Aster novi-belgii* 'Climax'

The original form of 'Climax' has beautiful lavender-blue flowers with a prominent golden disc. The most distinctive feature is the almost perfectly circular shape of the flower. However, another form that goes under the name 'Climax' is very similar but lacks the unusual neatness of the flowers. To differentiate between the two, the original form is often called 'Vicary Gibbs Climax'. Both are statuesque plants with strong sprays bearing prolific numbers of flowers.

ZONES 4–8
HABIT AND SIZE Vigorous spreading clumps produce strong upright sprays, 135 cm (53 in.) tall.
FLOWERS 5 cm (2 in.) across, in midautumn.
CULTIVATION Full sun, in good moisture-retentive garden soil. Space 60 cm (24 in.) apart. Divide every one to three years. Less susceptible to powdery mildew than many forms of *S. novi-belgii* due to *S. laeve* influences.
ORIGIN Bred by Edwin Beckett pre-1905. Beckett was head gardener to the Hon. Vicary Gibbs at Aldenham in Hertfordshire, United Kingdom. It was at Aldenham that the earliest collection of asters was created in the late nineteenth century.
LANDSCAPE AND DESIGN USES Best used towards the back of the border where it can form large clumps without smothering other plants. Combine with other stout herbaceous perennials or medium-sized shrubs.
SIMILAR PLANTS Two vigorous growers like 'Climax' are lavender-blue *S. laeve* 'Nightshade', 152 cm (60 in.) tall, and white *S. novi-belgii* 'Sam Banham', 150 cm (59 in.) tall.

Symphyotrichum novi-belgii 'Coombe Radiance'

New York aster, long leaf aster, traditional Michaelmas daisy
SYNONYMS *Aster novi-belgii* 'Coombe Radiance'

This is the one of the deepest-coloured late-flowering varieties. The bright purple-red flowers are a good size, borne on neat strong upright sprays, making this a valuable addition to the late-season garden.

ZONES 4–8

HABIT AND SIZE Compact clumps produce neat, strong sprays with upright branches, 102 cm (40 in.) tall.

FLOWERS 4.5 cm (1.75 in.) across, starting in early autumn.

CULTIVATION Full sun, in good moisture-retentive garden soil. Space 60 cm (24 in.) apart. Divide every one to three years. Susceptible to powdery mildew.

ORIGIN Raised in 1963. As with most of the varieties with the prefix "Coombe," the breeder was Ronald Watts.

LANDSCAPE AND DESIGN USES Ideally, this New York aster should be used as a border plant alongside other late-season perennials or compact shrubs.

SIMILAR PLANTS *Symphyotrichum novi-belgii* 'Twinkle' has purple-red flowers, 102 cm (40 in.) tall.

Symphyotrichum novi-belgii 'Daniela'

New York aster, long leaf aster, traditional Michaelmas daisy
SYNONYM *Aster novi-belgii* 'Daniela'

Starting in midautumn, good-sized deep lilac flowers cover low compact mounds of dark green foliage. 'Daniela' is valuable for its size and depth of colour, and is hard to improve upon when thriving.

ZONES 4–8
HABIT AND SIZE Branching sprays form low compact mounds, 20 cm (8 in.) tall.
FLOWERS 4.5 cm (1.75 in.) across, from midautumn.
CULTIVATION Full sun, in good moisture-retentive garden soil. Space 30 cm (12 in.) apart. Divide every one to two years. Susceptible to powdery mildew.
ORIGIN A German introduction bred by H. Klose in 1991.
LANDSCAPE AND DESIGN USES Although this New York aster can be used in containers it is best as a front-of-border plant. Here a large group can make a big impact and works well with other low-growing herbaceous plants or shrubs, particularly some of the smallest members of genus *Abelia* with bright golden foliage.
SIMILAR PLANTS Deep pink *S. novi-belgii* 'Rosenwichtel', 25 cm (10 in.) tall.

Symphyotrichum novi-belgii 'Davey's True Blue'

New York aster, long leaf aster, traditional Michaelmas daisy
SYNONYM *Aster novi-belgii* 'Davey's True Blue'

Superb large deep purple-blue flowers with a golden disc make this plant quite distinctive. As the name suggests it is one of the best deep "blue" asters available.

ZONES 4–8

HABIT AND SIZE Sturdy sprays, 120 cm (4 ft.) tall, from strong compact clumps.

FLOWERS 5 cm (2 in.) across, from midautumn.

CULTIVATION Full sun, in good moisture-retentive garden soil. Space 60 cm (24 in.) apart. Divide every one to two years. Susceptible to powdery mildew.

ORIGIN Bred by V. G. Davey in 1960.

LANDSCAPE AND DESIGN USES Most suited to the middle of the border where the clumps are small enough to not out-compete other plants, yet strong enough to hold their own. Towards the centre of the border, this New York aster provides a good colour break, combining well with yellows, reds, oranges, and the other warm shades largely found in autumnal colour schemes.

SIMILAR PLANTS Three cultivars of *S. novi-belgii*: 'Faith', 'Gurney Slade', and 'Harrison's Blue'. 'Faith' has deep lavender-blue flowers with a prominent yellow disc; its flowers closely resemble those of 'Davey's True Blue' but the plant is much more compact, 75 cm (30 in.) tall. 'Gurney Slade' has deep violet-blue, formally shaped double flowers, 90 cm (36 in.) tall. 'Harrison's Blue' has deep violet-blue flowers that are able to withstand inclement weather, 102 cm (40 in.) tall.

Symphyotrichum novi-belgii 'Dietgard'

New York aster, long leaf aster, traditional Michaelmas daisy
SYNONYM *Aster novi-belgii* 'Dietgard'

Very bushy low growth with fine mid-green leaves is covered with a mass of small bright pink flowers in midautumn. The number of flowers on each spray means that the clump appears smothered in a cloth of vibrant pink, rather than individual flowers.

ZONES 4–8
HABIT AND SIZE Strong compact clumps produce sprays that branch into neat mounds, 35 cm (14 in.) tall.
FLOWERS Less than 2.5 cm (1 in.) across, in midautumn.
CULTIVATION Full sun, in good moisture-retentive garden soil. Space 30 cm (12 in.) apart. Divide every one to two years. Susceptible to powdery mildew.
ORIGIN Unknown
LANDSCAPE AND DESIGN USES This plant works well as a specimen plant or as part of a group in the border. Like many of the dwarf New York asters, 'Dietgard' also makes a super container plant.
SIMILAR PLANTS The flowers of *S. novi-belgii* 'Kassel' are a vivid shade of light purple-red, 40 cm (16 in.) tall.

Symphyotrichum novi-belgii 'Elta'

New York aster, long leaf aster, traditional Michaelmas daisy
SYNONYM *Aster novi-belgii* 'Elta'

This plant is a member of a select group of New York aster cultivars that remain healthy and floriferous even when left undivided for a number of years. The bright purple-pink 'Elta' is also generally more disease resistant than many, making it a valuable addition for low-maintenance gardens.

ZONES 4–8
HABIT AND SIZE Compact clumps produce strong branching sprays, 90 cm (36 in.) tall.
FLOWERS 3.5 cm (1.5 in.) across, from early to midautumn.
CULTIVATION Full sun, in good moisture-retentive garden soil. Space 50 cm (20 in.) apart. Divide every one to four years. Sometimes susceptible to powdery mildew.
ORIGIN Unknown
LANDSCAPE AND DESIGN USES An excellent border plant requiring little maintenance, 'Elta' makes a very striking combination with plants that pick up its purple-pink colouration in either foliage or flower. For example, a particularly nice mix is the annual *Nicotiana mutabilis* grown behind 'Elta', where the *Nicotiana* flowers carry the colouration further back into the border. It would also make a great addition to more naturalistic planting schemes.
SIMILAR PLANTS *Symphyotrichum novi-belgii* 'Patricia Ballard' can also be left undivided for a number of years and has similar purple-pink flowers but more double.

Symphyotrichum novi-belgii 'Fellowship'

New York aster, long leaf aster, traditional Michaelmas daisy
SYNONYM *Aster novi-belgii* 'Fellowship'

If blowsy blooms are your thing then 'Fellowship' certainly fits the bill. Each pale pink petal is quilled creating a large well-shaped double flower with the appearance of being feathered.

ZONES 4–8
HABIT AND SIZE Small clumps produce strong sprays, 102 cm (40 in.) tall.
FLOWERS 5.5 cm (2.25 in.) across, from midautumn.
CULTIVATION Full sun, in good moisture-retentive garden soil. Space 50 cm (20 in.) apart. Divide every one to two years. Sometimes susceptible to powdery mildew.
ORIGIN This was introduced by Carlile in 1955 but was bred at Sandford's nurseries, which created some of the finest formal borders of New York asters and raised many excellent cultivars before being turned into a garden centre in the 1960s.

LANDSCAPE AND DESIGN USES As with most large-flowered New York asters, staking is advisable to ensure survival of the plant through autumn storms. For this reason, it is best positioned towards the centre of the border where any staking is easy to conceal. Try combining 'Fellowship' with deep purple flowers or purple tinted foliage to highlight the glorious size and softness of the flower. The sprays are also well suited to cutting.
SIMILAR PLANTS 'Coombe Margaret' has similar pale-pink double flowers but slightly smaller and earlier, 90 cm (36 in.) tall. 'Peace' also has similar flower size and shape but is lilac in colour, 90 cm (36 in.) tall; it was the forerunner to 'Fellowship'. 'Flamingo' has virtually identical if slightly smaller flowers but only reaches 80 cm (32 in.) tall.

Symphyotrichum novi-belgii 'Freya'

New York aster, long leaf aster, traditional Michaelmas daisy
SYNONYM *Aster novi-belgii* 'Freya'

This plant has distinctive large purple-pink flowers with quilled petals, which appear rather ragged when considered individually. However, the overall effect is opulent and quite desirable.

ZONES 4–8

HABIT AND SIZE Strong compact clumps produce upright sprays, 100 cm (39 in.) tall.

FLOWERS 5 cm (2 in.) across, from early to midautumn.

CULTIVATION Full sun, in good moisture-retentive garden soil. Space 50 cm (20 in.) apart. Divide every one to two years. Susceptible to powdery mildew.

ORIGIN Unknown

LANDSCAPE AND DESIGN USES Used most often to create swathes of rich colour in the centre of borders, this New York aster can also be used as a specimen plant to give a lift of colour as everything else is beginning to fade.

SIMILAR PLANTS *Symphyotrichum novi-belgii* 'Helen' has purple-pink double flowers; it is neater in form than 'Freya', 90 cm (36 in.) tall.

Symphyotrichum novi-belgii 'Gayborder Royal'

New York aster, long leaf aster, traditional Michaelmas daisy
SYNONYM *Aster novi-belgii* 'Gayborder Royal'

The flowers of 'Gayborder Royal' are without doubt some of the most distinctively shaped among the New York asters. Each small deep lilac-blue petal is neatly rounded at the tip, creating a perfectly shaped flower without any of the ragged edges seen in some other cultivars.

ZONES 4–8
HABIT AND SIZE Small clumps produce upright sprays, 70 cm (28 in.) tall.
FLOWERS 2.5 cm (1 in.) across, emerging earlier than some varieties, normally starting in early autumn.
CULTIVATION Full sun, in good moisture-retentive garden soil. Space 40 cm (16 in.) apart. Divide every one to two years. Susceptible to powdery mildew.

ORIGIN As the name suggests, it was raised at Gayborder Nurseries, United Kingdom, by Arthur Herbert Harrison during his ownership between the late 1920s and 1953.
LANDSCAPE AND DESIGN USES The form of the flower ensures that it stands out in an autumn border when planted as a group. The compact nature of the plant also allows it to be used as a container specimen; however, it is advisable to surround it with a lower growing plant due to a tendency to carry the flowers mainly at the top of the plant.
SIMILAR PLANTS The violet-blue flowers of *S. novi-belgii* 'Dora Chiswell' lack the finesse of 'Gayborder Royal' but the clumps are more compact, 45 cm (18 in.) tall.

Symphyotrichum novi-belgii 'Goliath'

New York aster, long leaf aster, traditional Michaelmas daisy
SYNONYM *Aster novi-belgii* 'Goliath'

It's easy to identify this New York aster from a distance: the large single lilac-pink flowers are held on short stems on upright sprays forming a definite pyramidal shape. The flowers are not the only large point; the entire plant has generous proportions from the broad tall sprays to the wide-toothed mid-green leaves, the latter indicating a *S. laeve* influence, and with it comes the associated improved resistance to powdery mildew.

ZONES 4–8
HABIT AND SIZE The pyramidal sprays from vigorous clumps are 130 cm (51 in.) tall.
FLOWERS 5.5 cm (2.25 in.) across, usually starting in midautumn.
CULTIVATION Full sun, in good moisture-retentive garden soil. Space 60 cm (24 in.) apart. Divide every one to four years. Shows an improved resistance to powdery mildew.
ORIGIN One of the varieties raised by Mrs. Thornely in her Devizes garden during the 1940s and 1950s.
LANDSCAPE AND DESIGN USES Being of a robust constitution and quite happy to be left undivided for a number of years, this selection can be used both in traditional borders or more naturalistic plantings where it can be allowed to form a grand swathe. The sprays are also well suited to use as cut flowers.
SIMILAR PLANTS *Symphyotrichum novi-belgii* 'Algar's Pride' closely resembles 'Goliath' but flowers earlier, 152 cm (5 ft.) tall, and *S. novi-belgii* 'Mammoth' is also very similar, but the flowers lack the pink tint, 130 cm (51 in.) tall.

Symphyotrichum novi-belgii 'Gurney Slade'

New York aster, long leaf aster, traditional Michaelmas daisy
SYNONYM *Aster novi-belgii* 'Gurney Slade'

The well-shaped flowers of this cultivar are the bold purple-blue and yellow combination often seen among New York asters. Although a common colour, the shade of this particular variety is incredibly vivid, improving the contrast made with the deep yellow disc. The sprays are well branched, creating an open bouquet of flowers.

ZONES 4–8
HABIT AND SIZE Compact clumps produce well-branched sprays at 90 cm (36 in.) tall.
FLOWERS 4.5 cm (1.75 in.) across, from early autumn onwards.
CULTIVATION Full sun, in good moisture-retentive garden soil. Space 50 cm (20 in.) apart. Divide every one to two years. Susceptible to powdery mildew.
ORIGIN This excellent cultivar was produced from Ernest Ballard's lines and raised by George Chiswell in 1966. The plant is named after a Somerset mining village.
LANDSCAPE AND DESIGN USES An ideal border plant best placed towards the front of the central section. This particular shade of purple-blue works well alongside a full palette of flower colours but particularly nice combinations can be created using bright reds, golden yellows, and deep purple-pinks. The sprays are also well shaped and suitable for cutting.
SIMILAR PLANTS *Symphyotrichum novi-belgii* 'Tovarich' has small, violet double flowers, 90 cm (36 in.) tall, and *S. novi-belgii* 'Harrison's Blue' has deep violet-blue flowers, 102 cm (40 in.) tall.

Symphyotrichum novi-belgii 'Heinz Richard'

New York aster, long leaf aster, traditional Michaelmas daisy
SYNONYM *Aster novi-belgii* 'Heinz Richard'

This is one of the showiest dwarf cultivars with surprisingly large flowers for such a diminutive plant. Vibrant purple-pink flowers smother the compact mounds of fine foliage.

ZONES 4–8

HABIT AND SIZE Well-branched sprays from vigorous clumps produce compact mounds to 30 cm (12 in.) tall.

FLOWERS 5 cm (2 in.) across, from early autumn onwards.

CULTIVATION Full sun, in good moisture-retentive garden soil. Space 30 cm (12 in.) apart. Divide every one to three years. Susceptible to powdery mildew.

ORIGIN Another fine introduction from the German breeder H. Klose in 1978.

LANDSCAPE AND DESIGN USES The mounding nature of the plant makes it an excellent specimen plant, perfect for the front of borders or containers. However, like many varieties it does look outstanding as a large group.

SIMILAR PLANTS *Symphyotrichum novi-belgii* 'Ilse Brensell' has lilac-pink flowers, mounds to 30 cm (12 in.) tall.

Symphyotrichum novi-belgii 'Helen Ballard'

New York aster, long leaf aster
SYNONYM *Aster novi-belgii* 'Helen Ballard'

This selection is one of the most striking with bright purple-red petals arranged formally around a golden yellow disc. These eye-catching flowers are held on well-shaped sprays over dark green foliage giving an extra intensity to the entire plant.

ZONES 4–8
HABIT AND SIZE Nicely shaped sprays from compact clumps to 102 cm (40 in.) tall.
FLOWERS 5 cm (2 in.) across, from early autumn onwards.
CULTIVATION Full sun, in good moisture-retentive garden soil. Space 50 cm (20 in.) apart. Divide every one to two years. Susceptible to powdery mildew.
ORIGIN Bred in 1962 at Old Court Nurseries, Colwall, by Percy Picton. Named for Ernest Ballard's daughter-in-law, the Helen Ballard of hellebore fame.
LANDSCAPE AND DESIGN USES Outstanding when planted as a group in the border. 'Helen Ballard' has such a depth of colour that you can use it to draw the eye in, and additionally, as a foil for paler plants.
SIMILAR PLANTS *Symphyotrichum novi-belgii* 'Freda Ballard' has bright purple-red flowers produced from early autumn on well-branched sprays to 90 cm (36 in.) tall. *Symphyotrichum novi-belgii* 'Janet Watts' produces purple-red flowers from early autumn on upright sprays to 102 cm (40 in.) tall.

Symphyotrichum novi-belgii 'Jean'

New York aster, long leaf aster, traditional Michaelmas daisy
SYNONYM *Aster novi-belgii* 'Jean'

The small violet flowers are attractive but not astounding. However, they are quite weatherproof, standing up well to endlessly wet days in autumn. It is also worth noting that the foliage is broad for a dwarf cultivar.

ZONES 4–8
HABIT AND SIZE Upright sprays from quite weak clumps grow to 45 cm (18 in.) tall.
FLOWERS 2.5 cm (1 in.) across, generally starting in midautumn.
CULTIVATION Full sun, in good moisture-retentive garden soil. Space 30 cm (12 in.) apart. Divide every one to two years. Susceptible to powdery mildew.
ORIGIN Bred by E. C. Simmonds and Sons around 1947.
LANDSCAPE AND DESIGN USES 'Jean' is not a very good specimen plant, as the clumps have no inclination to mound, so it is best used as a group, where the late flowers can prove invaluable to the autumn border.
SIMILAR PLANTS *Symphyotrichum novi-belgii* 'Gulliver' is a more refined plant than 'Jean' with pale heather-purple flowers, 45 cm (18 in.) tall.

Symphyotrichum novi-belgii 'Jenny'

New York aster, long leaf aster, traditional Michaelmas daisy
SYNONYM *Aster novi-belgii* 'Jenny'

This excellent compact cultivar with small bushy sprays is probably the most intensely coloured dwarf New York aster. The bright purple-red, double flowers are large for the compact habit of the plant.

ZONES 4–8
HABIT AND SIZE Small clumps produce tight bushy sprays to 40 cm (16 in.) tall.
FLOWERS 5 cm (2 in.) across, from early autumn onwards.
CULTIVATION Full sun, in good moisture-retentive garden soil. Space 30 cm (12 in.) apart. Divide every one to two years. Susceptible to powdery mildew.

ORIGIN Whether this cultivar is correctly named or not is the subject of some debate. Harrison bred the original plant before 1927, but the current plant did not appear until the mid 1960s. In 1966 'Royal Ruby' was released by Blooms of Bressingham and is identical to the current 'Jenny'; the only difference we have ever noted is that 'Jenny' has strongly scented roots and 'Royal Ruby' does not, which is not a useful identification feature.
LANDSCAPE AND DESIGN USES Either this cultivar or 'Royal Ruby' make superb front-of-border or container plants. 'Jenny' is also among the few compact varieties that work well as a cut flower because it has a reasonable length of stem and the sprays are not too large.
SIMILAR PLANTS As discussed, *S. novi-belgii* 'Royal Ruby' is virtually identical to 'Jenny, while *S. novi-belgii* 'Terry's Pride' has large, rich purple-pink, double flowers with a similar habit and flower form as 'Jenny', growing to 45 cm (18 in.) tall.

Symphyotrichum novi-belgii 'Lady in Blue'

New York aster, long leaf aster, traditional Michaelmas daisy
SYNONYM *Aster novi-belgii* 'Lady in Blue'

Low mounds of narrow foliage are smothered in double lavender-blue flowers. The flowers show no disc for a remarkably long time creating an unusual and very effective show.

ZONES 4–8
HABIT AND SIZE Strong but compact clumps produce bushy sprays to 30 cm (12 in.) tall.
FLOWERS 4.5 cm (1.75 in.) across, from early autumn onwards.
CULTIVATION Full sun, in good moisture-retentive garden soil. Space 30 cm (12 in.) apart. Divide every one to two years. Susceptible to powdery mildew.
ORIGIN A fine introduction by Carlile in 1955 from the renowned breeder Amos Perry.
LANDSCAPE AND DESIGN USES 'Lady in Blue' can make a nice specimen plant but it is often more effective when used in a group. Excellent as a container plant.
SIMILAR PLANTS *Symphyotrichum novi-belgii* 'Rose Bonnet' is a superb dwarf cultivar with double pale lilac-pink flowers in midautumn, 35 cm (14 in.) tall. The flower shape and growth habit are very close to that of 'Lady in Blue'.

Symphyotrichum novi-belgii 'Lassie'

New York aster, long leaf aster, traditional Michaelmas daisy

SYNONYM *Aster novi-belgii* 'Lassie'

The soft, gentle appearance of the petals and the reliability of this plant are fittingly reminiscent of the famous character, Lassie the dog. The large pale pink flowers are produced en masse over a remarkably long period. The growth and form of the plant as a whole are attractive and flowers are often produced down the entire stem, although sometimes the lower flowers are single.

ZONES 4–8

HABIT AND SIZE Strong clumps bear sturdy sprays reaching 102 cm (40 in.) tall.

FLOWERS 5 cm (2 in.) across, from early to midautumn.

CULTIVATION Full sun, in good moisture-retentive garden soil. Space 60 cm (24 in.) apart. Divide every one to three years. Susceptible to powdery mildew.

ORIGIN Carlile introduced this cultivar around 1955 from Sandford's nurseries.

LANDSCAPE AND DESIGN USES An excellent cultivar for the border. The pale colouring means that it works very effectively with a dark foil. An unusual combination can be achieved by growing 'Lassie' in front of a sun-tolerant *Acer* cultivar such as *A. palmatum* 'Bloodgood'.

SIMILAR PLANTS *Symphyotrichum novi-belgii* 'Fellowship' has pale pink double flowers, and *S. novi-belgii* 'Timsbury' has very pale pink formal double flowers. Both grow to 102 cm (40 in.) tall like 'Lassie'.

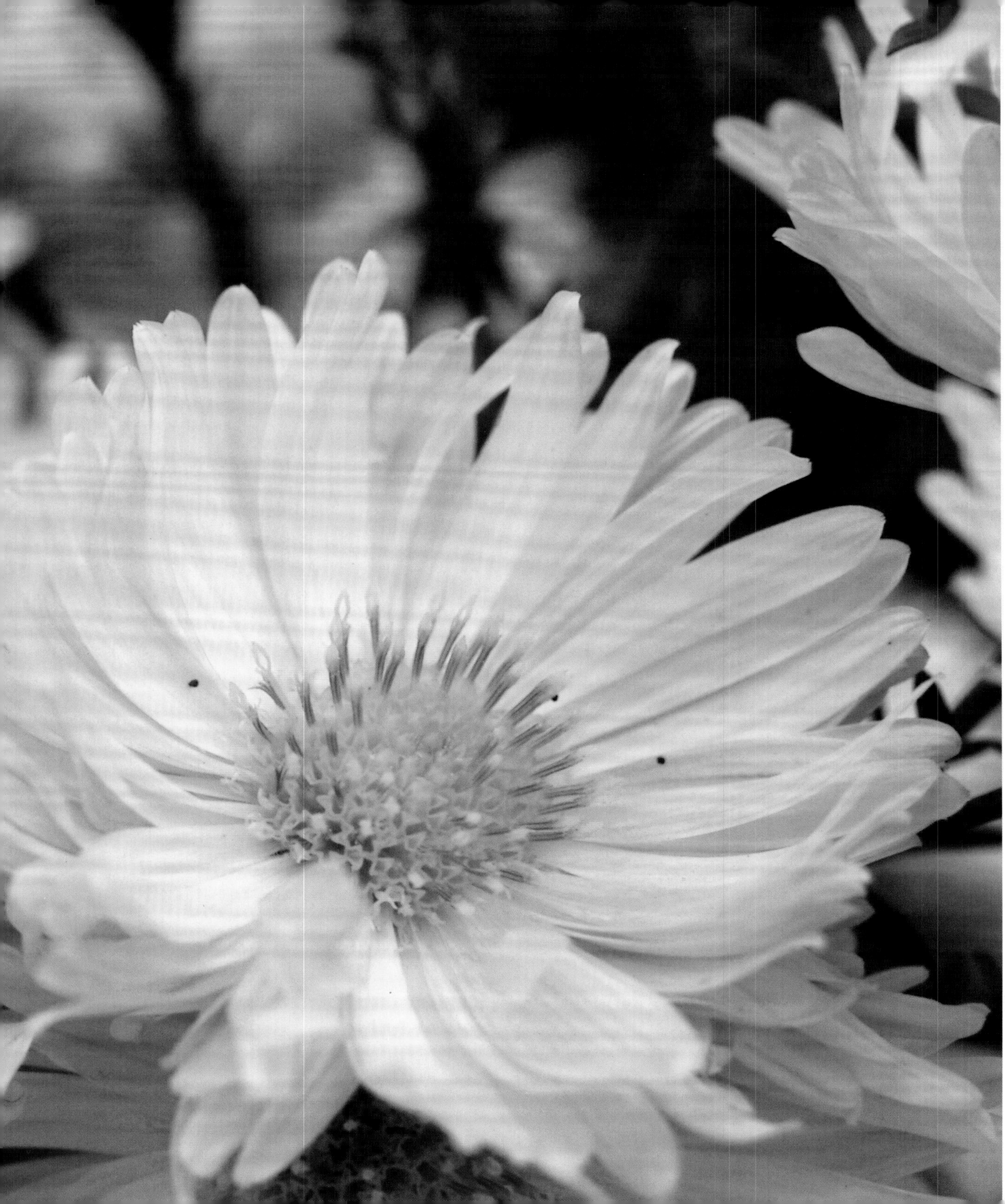

Symphyotrichum novi-belgii 'Lisa Dawn'

New York aster, long leaf aster, traditional Michaelmas daisy

SYNONYM *Aster novi-belgii* 'Lisa Dawn'

This is one of the earliest-flowering varieties of New York aster. Often in late summer the bushy green sprays will transform into bouquets of large mulberry-red flowers with bright golden yellow discs.

ZONES 4–8

HABIT AND SIZE Strong clumps produce bushy sprays to 45 cm (18 in.) tall.

FLOWERS 4.5 cm (1.75 in.) across, usually starting in late summer and continuing for several weeks (deadheading is advisable).

CULTIVATION Full sun, in good moisture-retentive garden soil. Space 40 cm (16 in.) apart. Divide every one to two years. Susceptible to powdery mildew.

ORIGIN Unknown

LANDSCAPE AND DESIGN USES Suitable either as a border plant or for growing in a container. Mixes well with blue and pale yellow herbaceous perennials, catching the late summer varieties as well as the autumn ones.

SIMILAR PLANTS *Symphyotrichum novi-belgii* 'Rufus' has bright purple-red flowers similar in colour to 'Lisa Dawn' but is taller and bushier, 90 cm (36 in.).

Symphyotrichum novi-belgii 'Little Man in Blue'

New York aster, long leaf aster, traditional Michaelmas daisy
SYNONYM *Aster novi-belgii* 'Little Man in Blue'

One of the most reliably floriferous varieties available. The flowers create a sea of lavender-blue over broad foliaged mounds. The floriferous nature of this plant is matched by its vigorous growth.

ZONES 4–8
HABIT AND SIZE Vigorous clumps produce compact mounds of bushy sprays reaching 35 cm (14 in.) tall.
FLOWERS 4.5 cm (1.75 in.) across, from early autumn onwards.
CULTIVATION Full sun, in good moisture-retentive garden soil. Space 40 cm (16 in.) apart. Divide every one to three years. Susceptible to powdery mildew.
ORIGIN One of Ernest Ballard's few dwarf cultivars. Raised in 1935.
LANDSCAPE AND DESIGN USES Ideal as a container plant or in the border. When used in the border it is a useful aid to repeating colours down the length of the border forming compact groups. However, it lends itself to a more exuberant approach where large groups can be created with relatively few plants.
SIMILAR PLANTS *Symphyotrichum novi-belgii* 'Little Boy Blue' has violet-blue flowers; habit is similar to 'Little Man in Blue' but slightly taller at 40 cm (16 in.). *Symphyotrichum novi-belgii* 'Little Pink Beauty' is similarly floriferous and vigorous but with bright mauve-pink flowers, 35 cm (14 in.) tall.

Symphyotrichum novi-belgii 'Little Pink Beauty'

New York aster, long leaf aster, traditional Michaelmas daisy
SYNONYM *Aster novi-belgii* 'Little Pink Beauty'

Among the strongest and most reliably free-flowering compact varieties available. Beginning in early autumn, good-sized bright mauve-pink flowers smother the mounds of broad mid-green foliage.

ZONES 4–8
HABIT AND SIZE Vigorous clumps produce low mounds of branching sprays, 35 cm (14 in.) tall.
FLOWERS 3.5 cm (1.5 in.) across, from early to midautumn.
CULTIVATION Full sun, in good moisture-retentive garden soil. Space 40 cm (16 in.) apart. Divide every one to three years. Susceptible to powdery mildew.

ORIGIN Bred in the United Kingdom by G. Shepheard-Walwyn in 1959.
LANDSCAPE AND DESIGN USES Each individual plant makes a nice shape and flowers well, which makes it suitable for use as a specimen plant in the border or container. However, a generous group of this New York aster will be a true asset to any border.
SIMILAR PLANTS *Symphyotrichum novi-belgii* 'Little Pink Lady' has soft mauve-pink flowers in mid-autumn, 40 cm (16 in.) tall, and *S. novi-belgii* 'Little Pink Pyramid' produces purple-pink flowers from early autumn, 45 cm (18 in.) tall.

Symphyotrichum novi-belgii 'Marie Ballard'

New York aster, long leaf aster, traditional Michaelmas daisy
SYNONYM *Aster novi-belgii* 'Marie Ballard'

It would not be an exaggeration to call 'Marie Ballard' the best of the double-flowered varieties ever to have been bred. As testament to this, many doppelgangers have appeared over the years. The perfectly shaped double flowers have about 250 petals each, and last for a long time with the disc only becoming visible as the flower ages and the central petals unfurl. Although lavender-blue, the flowers are such a shade as to be considered true blue (or at least as near it as New York asters are ever likely to achieve).

ZONES 4–8
HABIT AND SIZE Flowering branches are bunched at the top of the 102 cm (40 in.) tall sprays, from strong compact clumps.
FLOWERS 4.5 cm (1.75 in.) across, from early to midautumn.
CULTIVATION Full sun, in good moisture-retentive garden soil. Space 60 cm (24 in.) apart. Divide every one to three years, though it can do surprisingly well if left for two or three years undivided. Susceptible to powdery mildew.
ORIGIN This was one of Ernest Ballard's last cultivars, bred just before his death and named for his second wife. It was introduced in 1955 by Percy Picton, then manager of Old Court Nurseries.
LANDSCAPE AND DESIGN USES This plant is best suited to a centre-of-the-border position, near enough to the front that the beauty of the individual flowers can be appreciated, but far enough back that the staking is invisible. The sprays are also suitable for cutting.
SIMILAR PLANTS *Symphyotrichum novi-belgii* 'Madge Cato' and *S. novi-belgii* 'Melbourne Magnet' produce lilac-blue, formal double flowers ('Madge Cato' is paler in colour) and are 90 cm (36 in.) tall.

Symphyotrichum novi-belgii 'Mount Everest'

New York aster, long leaf aster, traditional Michaelmas daisy

SYNONYM *Aster novi-belgii* 'Mount Everest'

Wonderful pyramidal sprays of white daisies. The flowers of 'Mount Everest' are large enough to stand out, without being so big that bad weather causes excessive damage.

ZONES 5–8

HABIT AND SIZE Strong spreading clumps with sturdy stems to 180 cm (6 ft.) tall.

FLOWERS 5 cm (2 in.) across, opening from early autumn onwards.

CULTIVATION Full sun, in good moisture-retentive garden soil. Space 60 cm (24 in.) apart. Divide every one to three years. Powdery mildew can be a problem.

ORIGIN Bred at Old Court Nurseries in Colwall before 1930 by Ernest Ballard.

LANDSCAPE AND DESIGN USES Most definitely a candidate for the back of the border where the grand pyramidal sprays can give a lift to the late-season colour of your garden. The sprays are also ideal cut flowers.

SIMILAR PLANTS *Symphyotrichum novi-belgii* 'Albanian', 102 cm (40 in.) tall, and *S. novi-belgii* 'Blandie', 90 cm (36 in.) tall, both produce white flowers from cream buds.

Symphyotrichum novi-belgii 'Norman's Jubilee'

New York aster, long leaf aster, traditional Michaelmas daisy
SYNONYM *Aster novi-belgii* 'Norman's Jubilee'

As one of the earliest New York asters to flower, the pale pink double blooms are always distinctive, and are large for the compact habit of this plant.

ZONES 4–8

HABIT AND SIZE Compact clumps produce well-branched upright sprays to 40 cm (16 in.) tall.

FLOWERS 4.5 cm (1.75 in.) across, produced during early autumn.

CULTIVATION Full sun, in good moisture-retentive garden soil. Space 40 cm (16 in.) apart. Divide every one to two years. Susceptible to powdery mildew.

ORIGIN Bred in 1985 by Bee's nurseries based in Cheshire, United Kingdom.

LANDSCAPE AND DESIGN USES Best used in groups rather than as specimen plants, this pretty, little aster is ideal for the front of the border.

SIMILAR PLANTS *Symphyotrichum novi-belgii* 'Chatterbox' is virtually identical to 'Norman's Jubilee' but flowers about two weeks later, 40 cm (16 in.) tall. *Symphyotrichum novi-belgii* 'Autumn Days' has similar large pale-pink flowers produced in early autumn, upright sprays to 60 cm (24 in.) tall.

Symphyotrichum novi-belgii 'Patricia Ballard'

New York aster, long leaf aster, traditional Michaelmas daisy

SYNONYM *Aster novi-belgii* 'Patricia Ballard'

Distinctive bright purple-pink double flowers are freely produced. Reliable flowering on relatively compact growth is achieved even if this plant is left undivided for a number of years.

ZONES 4–8

HABIT AND SIZE Vigorous clumps with strong well-branched sprays up to 90 cm (36 in.) tall.

FLOWERS 4.5 cm (1.75 in.) across, produced from early autumn.

CULTIVATION Full sun, in good moisture-retentive garden soil. Space 60 cm (24 in.) apart. Divide every one to four years. Susceptible to powdery mildew.

ORIGIN Bred at Old Court Nurseries in Colwall by Percy Picton in 1957.

LANDSCAPE AND DESIGN USES Probably due to its vigour, this aster has proved popular as a garden plant on both sides of the Atlantic. Given the room 'Patricia Ballard' will make a medium-sized clump and often flowers right down the plant. This means it can be used effectively at the front of the border instead of being relegated to the depths. The colour is bold without being overpowering so it is easy to use alongside other late-season plants be they perennials, annuals, or shrubs.

SIMILAR PLANTS *Symphyotrichum novi-belgii* 'Lady Frances' has deep mauve-pink, double flowers, produced a few weeks later than 'Patricia Ballard' on upright growth of 100 cm (40 in.). *Symphyotrichum novi-belgii* 'Pamela' is more of a mauve-pink in colour, but like 'Patricia Ballard', the double flowers are produced from early autumn, 90 cm (36 in.) tall.

Symphyotrichum novi-belgii 'Remembrance'

New York aster, long leaf aster, traditional Michaelmas daisy
SYNONYM *Aster novi-belgii* 'Remembrance'

When the astonishingly large lavender-blue flowers with pale yellow centres cover the plant, it makes a lasting impression, particularly because this usually occurs late in the season.

ZONES 4–8
HABIT AND SIZE Strong clumps produce sturdy sprays reaching 50 cm (20 in.) in height, forming mounds.
FLOWERS 5 cm (2 in.) across, produced from midautumn.
CULTIVATION Full sun, in good moisture-retentive garden soil. Space 60 cm (24 in.) apart. Divide every one to three years. Susceptible to powdery mildew.

ORIGIN This cultivar was bred by H. V. Vokes before 1935 as part of his work producing dwarf cultivars for the War Graves Commission.
LANDSCAPE AND DESIGN USES Even a single plant of this aster will make a lovely show with masses of flowers. This makes it a good choice for containerized growing or for small borders. However, the effect of one plant is only enhanced when planted as a group so it is equally suitable for large borders in both formal and informal settings.
SIMILAR PLANTS *Symphyotrichum novi-belgii* 'Blue Bouquet' has lavender-blue flowers, smaller than 'Remembrance', borne in midautumn, 40 cm (16 in.) tall. *Symphyotrichum novi-belgii* 'Cecily' is a very pale mauve-blue, 50 cm (20 in.) tall.

Symphyotrichum novi-belgii 'Rosebud'

New York aster, long leaf aster, traditional Michaelmas daisy
SYNONYM *Aster novi-belgii* 'Rosebud'

Masses of distinctive button-like flowers cover dark green very compact growth. The rose-pink petals are produced in such quantity that they usually conceal the disc for the majority of the life of each small flower.

ZONES 4–8

HABIT AND SIZE The sprays are short but sturdy reaching 30 cm (12 in.) in height. The clumps tend to be small and rather weak.

FLOWERS 2.5 cm (1 in.) across, produced from early autumn.

CULTIVATION Full sun, in good moisture-retentive garden soil. Space 30 cm (12 in.) apart. Divide every one to two years. Susceptible to powdery mildew.

ORIGIN Ernest Ballard had little interest in breeding dwarf asters, but you can clearly see his influence in this pre-1950 diminutive double-flowered plant.

LANDSCAPE AND DESIGN USES To appreciate this New York aster fully, it is advisable to grow it as a group at the front of a border where it will not be overcrowded. Also suitable for the larger rock garden if you are not an alpine purist. Good in containers.

SIMILAR PLANTS *Symphyotrichum novi-belgii* 'Snow Sprite' produces large, double white flowers on growth very similar to 'Rosebud'.

Symphyotrichum novi-belgii 'Rosenwichtel'

New York aster, long leaf aster, traditional Michaelmas daisy
SYNONYM *Aster novi-belgii* 'Rosenwichtel'

This may be one of the most diminutive varieties on the market, yet the deep pink flowers are still a good size and freely produced, covering the low mounds of broad dark green foliage.

ZONES 4–8

HABIT AND SIZE The clumps are vigorous and will spread producing numerous sturdy sprays to 25 cm (10 in.) high, branching into low mounds.

FLOWERS 3.5 cm (1.5 in.) across, produced from midautumn.

CULTIVATION Full sun, in good moisture-retentive garden soil. Space 40 cm (16 in.) apart. Divide every one to three years. Susceptible to powdery mildew.

ORIGIN Bred in 1969 by Zur Linden in Germany.

LANDSCAPE AND DESIGN USES Both compact and floriferous, this aster is suitable for container growth as well as the front of the border. It is also possible to use it along the lower regions of rock gardens.

SIMILAR PLANTS *Symphyotrichum novi-belgii* 'Daniela' has lilac double flowers and grows 20 cm (8 in.) tall.

Symphyotrichum novi-belgii 'Saint Egwyn'

New York aster, long leaf aster, traditional Michaelmas daisy
SYNONYM *Aster novi-belgii* 'Saint Egwyn'

The defining feature of this dainty New York aster is the mass of small pale purple-pink flowers that are held in numerous panicles. Flowering sprays grow at varying heights with this aster thereby creating a large dome of flowers.

ZONES 4–8
HABIT AND SIZE Vigorous clumps produce sturdy sprays with open branches to 102 cm (40 in.) in height.
FLOWERS 2.5 cm (1 in.) across, produced from early autumn.
CULTIVATION Full sun, in good moisture-retentive garden soil. Space 60 cm (24 in.) apart. Divide every two to four years. Although susceptible to powdery mildew, it is more resistant than many New York asters.
ORIGIN A very old cultivar, 'Saint Egwyn' was introduced before 1907.

LANDSCAPE AND DESIGN USES Attractive and floriferous even when left undivided for a number of years, this New York aster lends itself to use in both formal borders and more informal settings where it can be allowed to make clouds of pale purple-pink in the autumn. Eye-catching combinations can be achieved by planting 'Saint Egwyn' next to large-flowered varieties such as *Aster ×frikartii* 'Mönch', using the soft colours to complement each other and allowing the textural difference to do the talking. The sprays of 'Saint Egwyn' are also strong enough and with sufficient form to serve as cut flowers.
SIMILAR PLANTS *Symphyotrichum novi-belgii* 'Dazzler' has bright purple-pink flowers closely resembling 'Saint Egwyn' but is weaker in growth and with a less floriferous habit, 90 cm (36 in.) in height.

Symphyotrichum novi-belgii 'Sam Banham'

New York aster, long leaf aster, traditional Michaelmas daisy
SYNONYM *Aster novi-belgii* 'Sam Banham'

Something about the look of this desirable aster is undeniably romantic, particularly in autumn's low evening and morning light. Strong distinctive sprays form pyramids of large, white, well-shaped flowers.

ZONES 4–8
HABIT AND SIZE The clumps are vigorous and will spread, producing numerous upright but well-branched sprays to 150 cm (59 in.) high, branching into tall pyramids.
FLOWERS 5 cm (2 in.) across, lasting well into midautumn.
CULTIVATION Full sun. Will tolerate most soils, but prefers moisture-retentive garden soil. Space 60 cm (24 in.) apart. Divide every two to four years. Susceptible to powdery mildew, but rarely troubled.
ORIGIN Unknown
LANDSCAPE AND DESIGN USES This architectural plant is ideal in informal settings alongside perennial grasses, small to medium shrubs, or other herbaceous plants. It also lends itself to a cottage garden setting where it can produce billowing masses of white to blend with the other contenders for space. However, if regularly divided it will behave sufficiently well to be of use in more formal settings. Like many other New York asters, it is also ideal as a cut flower.
SIMILAR PLANTS *Symphyotrichum novi-belgii* 'Steinebrück' also has white flowers and strong growth to 120 cm (4 ft.) tall, but lacks some of the soft charm exhibited by 'Sam Banham'.

Symphyotrichum novi-belgii 'Sandford's White Swan'

New York aster, long leaf aster, traditional Michaelmas daisy

SYNONYM *Aster novi-belgii* 'Sandford's White Swan'

Fabulous double flowers emerge from pink-tinted buds. When open they are a soft pure white, the flowers almost appearing feather-like. As they age, the petals become tinged with purple-pink leading to the plant having an attractive bi-coloured appearance.

ZONES 4–8

HABIT AND SIZE Strong clumps produce sturdy upright growth, 102 cm (40 in.) tall, with numerous side shoots that can flower given space.

FLOWERS 4.5 cm (1.75 in.) across, from early autumn.

CULTIVATION Full sun, in good moisture-retentive garden soil. Space 60 cm (24 in.) apart. Divide every one to three years. Susceptible to powdery mildew.

ORIGIN Not surprisingly, 'Sandford's White Swan' was bred by Sandford's Nurseries. Located at Barton Mills, on the Cambridgeshire–Suffolk border in the United Kingdom, Sandford's devoted several acres to creating magnificent formal borders of New York asters. A number of cultivars were raised here before Sandford's became a garden centre in the 1960s.

LANDSCAPE AND DESIGN USES Given room, this cultivar is capable of producing flowers right down the plant, making it an ideal choice for a central position in more open borders or even towards the front. Create a striking combination by planting it adjacent to *S. novi-belgii* 'Purple Dome' or 'Pride of Colwall', where the white is really brought to life by the deep heather-purple in the other cultivars.

SIMILAR PLANTS *Symphyotrichum novi-belgii* 'Boningale White' produces white, semi-double flowers and grows 90 cm (36 in.) tall. *Symphyotrichum novi-belgii* 'Timsbury' has very pale pink, double flowers with a habit that closely resembles 'Sandford's White Swan'.

Symphyotrichum novi-belgii 'Schöne von Dietlikon'

New York aster, long leaf aster, traditional Michaelmas daisy
SYNONYM *Aster novi-belgii* 'Schöne von Dietlikon'

The two points that make this cultivar stand out from the crowd are the quantity and quality of the flowers. Sturdy, well-shaped sprays are smothered in a mass of perfectly shaped single violet-blue flowers with prominent yellow discs.

ZONES 4–8
HABIT AND SIZE The clumps are strong but stay relatively compact. Numerous sturdy sprays with short branches reach a height of 102 cm (40 in.).
FLOWERS 3.5 cm (1.5 in.) across, borne from early autumn.
CULTIVATION Full sun, in good moisture-retentive garden soil. Space 60 cm (24 in.) apart. Divide every two to three years. Susceptible to powdery mildew, but rarely troubled.
ORIGIN One of the least known members of the Frikart stable in Switzerland where the famous *Aster* ×*frikartii* originated.
LANDSCAPE AND DESIGN USES 'Schöne von Dietlikon' is fantastic in the border, putting on a superb show over a long period. The clumps are strong enough to mix with grasses, but not too vigorous to be used in formal or small borders. The sprays are also ideal for cutting.
SIMILAR PLANTS *Symphyotrichum novi-belgii* 'Faith' is a deep lavender-blue, 75 cm (30 in.) tall.

Symphyotrichum novi-belgii 'Sheena'

New York aster, long leaf aster, traditional Michaelmas daisy
SYNONYM *Aster novi-belgii* 'Sheena'

Well-shaped, large, double flowers of an unusual clean shade of deep pink are the defining features of this attractive New York aster.

ZONES 4–8
HABIT AND SIZE Sturdy sprays with open branches from moderately vigorous clumps, 102 cm (40 in.) in height.
FLOWERS 5 cm (2 in.) across, borne from early to midautumn.
CULTIVATION Full sun, in good moisture-retentive garden soil. Space 60 cm (24 in.) apart. Divide every one to two years. Susceptible to powdery mildew.

ORIGIN Bred by Ronald Watts in the United Kingdom around 1969.
LANDSCAPE AND DESIGN USES The unusual colour of the flowers makes 'Sheena' a good addition to the autumnal border. It is also an excellent cut flower thanks to the high quality of the blooms and nice shape of the sprays.
SIMILAR PLANTS *Symphyotrichum novi-belgii* 'Elizabeth Hutton' has the same unusual shade of pink as 'Sheena', but with smaller flowers, 90 cm (36 in.) tall. *Symphyotrichum novi-belgii* 'Irene', 102 cm (40 in.) tall, is similar in colour to both 'Elizabeth Hutton' and 'Sheena' but with smaller single flowers. 'Irene' was forerunner to 'Elizabeth Hutton'.

Symphyotrichum novi-belgii 'Sophia'

New York aster, long leaf aster, traditional Michaelmas daisy
SYNONYM *Aster novi-belgii* 'Sophia'

The bold colour and neat shape of the flowers ensure that this cultivar stands out from the crowd. The vibrant deep purple-pink flowers are small and double, with the petals neatly packed around a small yellow disc.

ZONES 4–8
HABIT AND SIZE Strong compact clumps and sprays, reaching 90 cm (36 in.) in height.
FLOWERS 2.5 cm (1 in.) across, borne from early to midautumn.
CULTIVATION Full sun, in good moisture-retentive garden soil. Space 60 cm (24 in.) apart. Divide every one to two years. Susceptible to powdery mildew.
ORIGIN Bred by Ronald Watts in the United Kingdom in 1968.
LANDSCAPE AND DESIGN USES Best used towards the middle of a border where the flowers can be appreciated and the lower stems ignored. Eye-catching combinations can be made using this aster; even just setting it against strong green foliage is very effective. Lovely as a cut flower.
SIMILAR PLANTS *Symphyotrichum novi-belgii* 'Beechwood Charm' has deep pink single flowers that are smaller and less vivid than 'Sophia', 100 cm (39 in.) tall.

Symphyotrichum novi-belgii 'Thundercloud'

New York aster, long leaf aster, traditional Michaelmas daisy
SYNONYM *Aster novi-belgii* 'Thundercloud'

A most unusual shade of deep heather-purple, closely resembling, as the name suggests, angry clouds gathering in the gloaming. The flowers are double but quite small and appear late in the season.

ZONES 4–8
HABIT AND SIZE Compact clumps produce upright sprays to 120 cm (4 ft.) in height.
FLOWERS 3.5 cm (1.5 in.) across, in midautumn.
CULTIVATION Full sun, in good moisture-retentive garden soil. Space 60 cm (24 in.) apart. Divide every one to two years. Susceptible to powdery mildew.
ORIGIN Unknown
LANDSCAPE AND DESIGN USES Most suitable for mixing with paler colours where the dark colour of 'Thundercloud' can provide much-needed depth. The habit of growth lends it very well to naturalistic style of planting and it makes a lovely cut flower, particularly when used with autumnal foliage.
SIMILAR PLANTS *Symphyotrichum novi-belgii* 'Dusky Maid' has large, double, deep dusky purple-pink flowers, 120 cm (4 ft.) tall. *Symphyotrichum novi-belgii* 'Pride of Colwall' has similarly sized and shaped heather-purple flowers as 'Thundercloud', 90 cm (36 in.) tall.

Symphyotrichum novi-belgii 'Trudi Ann'

New York aster, long leaf aster, traditional Michaelmas daisy
SYNONYM *Aster novi-belgii* 'Trudi Ann'

The most striking feature of this cultivar is the size of flower in contrast to the compact nature of the plant. These huge deep heather-purple flowers smother the large mounds of mid-green foliage.

ZONES 4–8
HABIT AND SIZE The clumps are small but bear numerous sprays that branch to form mounds to 40 cm (16 in.) in height.
FLOWERS 5 cm (2 in.) across, borne from early autumn.
CULTIVATION Full sun, in good moisture-retentive garden soil. Space 40 cm (16 in.) apart. Divide every one to three years. Susceptible to powdery mildew, but not as prone as some plants.
ORIGIN Unknown
LANDSCAPE AND DESIGN USES The compact nature of this cultivar makes it ideal for the front of border. It also makes a suitable candidate for growing in containers, particularly since the sprays form mounds meaning that it will happily fill a container with no need for companion plants.
SIMILAR PLANTS Like 'Trudi Ann', *S. novi-belgii* 'Blue Lagoon' and *S. novi-belgii* 'Mauve Magic' both produce flowers that are large for the size of the plant. 'Blue Lagoon' has deep violet-blue flowers on a plant that grows 45 cm (18 in.) tall; deep dusky purple-pink 'Mauve Magic' reaches 40 cm (16 in.) in height.

Symphyotrichum novi-belgii 'White Wings'

New York aster, long leaf aster, traditional Michaelmas daisy
SYNONYM *Aster novi-belgii* 'White Wings'

The flowers on this cultivar are very large and incredibly striking. The petals are numerous and narrow, of an exceptionally clean white. These long petals surround a small disc, giving the flower a high petal to disc ratio. 'White Wings' is also noted for having a long flowering season.

ZONES 4–8

HABIT AND SIZE Strong clumps produce sturdy sprays with rather lax branches, reaching 90 cm (36 in.) in height.

FLOWERS 5 cm (2 in.) across, produced from early autumn.

CULTIVATION Full sun, in good moisture-retentive garden soil. Space 60 cm (24 in.) apart. Divide every one to two years. Susceptible to powdery mildew.

ORIGIN Unknown

LANDSCAPE AND DESIGN USES This striking cultivar makes an excellent cut flower. In the border, it stands out and can look superb, although it does need to be well supported. If planted with something dark behind it, such as *Ageratina altissima* 'Chocolate', 'White Wings' is simply astonishing.

SIMILAR PLANTS *Symphyotrichum novi-belgii* 'Blue Radiance' is lavender-blue rather than white, but the two cultivars bear a close resemblance in terms of flower form, height, and the need for support.

Symphyotrichum novi-belgii 'Winston S. Churchill'

New York aster, long leaf aster, traditional Michaelmas daisy
SYNONYM *Aster novi-belgii* 'Winston S. Churchill'

This cultivar is exceptional because of the beautiful shade of purple-red flowers and the overall shape of the plant. The sprays have numerous bushy branches, which are then covered in the bright flowers.

ZONES 4–8
HABIT AND SIZE Compact clumps produce sturdy bushy sprays to 80 cm (32 in.) tall.
FLOWERS 4.5 cm (1.75 in.) across, produced from early autumn.
CULTIVATION Full sun, in good moisture-retentive garden soil. Space 60 cm (24 in.) apart. Divide every one to two years. Susceptible to powdery mildew.
ORIGIN Bred by A. H. Harrison and then introduced by Gayborder Nurseries around 1950, this plant was of course an instant hit for the name if nothing else. It has since remained a firm favourite with the public.

LANDSCAPE AND DESIGN USES Given a dry autumn and having managed to avoid mildew 'Winston S. Churchill' is hard to beat in the border or large container. A superb combination can be created using a pale blue such as *S. novi-belgii* 'Cecily', or with soft-structured plants such as *Stipa tenuissima*.
SIMILAR PLANTS *Symphyotrichum novi-belgii* 'Peter Chiswell' has bright purple-red flowers that are more double than 'Winston S. Churchill', 75 cm (30 in.) tall. The bright purple-red flowers of *S. novi-belgii* 'Rufus' closely resemble those of 'Winston S. Churchill' but the plant has much stronger growth, is more resistant to mildew, and lacks the distinctive form, 90 cm (36 in.) in height.

Symphyotrichum oblongifolium 'Fanny's Aster'

Aromatic aster
SYNONYM *Aster oblongifolius* 'Fanny's Aster'

This is one of the toughest plants within the group we are calling asters. Like the type species, 'Fanny's Aster' is both heat hardy and cold hardy. Over a long period, masses of violet flowers cover the strong bushy growth. The foliage is hairy and lightly scented when crushed, as the common name suggests.

ZONES 4–9
HABIT AND SIZE The stems are woody with many horizontal stiff branches up their length borne from compact clumps. This plant reaches between 60 cm (24 in.) and 90 cm (36 in.) in height, and in the United States will happily take being cut back in the early summer to reduce the height.
FLOWERS 1.8 cm (0.75 in.) across, often appearing from late summer to midautumn in the United Kingdom, and continuing through late autumn in areas of the United States.
CULTIVATION Full sun, in any reasonable garden soil. Space at least 60 cm (24 in.) apart. Divide every four to five years. Mildew free.

ORIGIN Nancy Goodwin at Montrose, her historic garden in North Carolina, introduced this cultivar.
LANDSCAPE AND DESIGN USES Due to the woody nature and bushy habit of this cultivar, 'Fanny's Aster' makes a super specimen plant, combining well with small shrubs or trees such as miniature pines. It also makes a wonderful border plant where the flowers can be shown to perfection alongside the earlier flowering varieties of hardy chrysanthemum, particularly those with button-like flowers such as 'Nantyderry Sunshine'.
SIMILAR PLANTS *Symphyotrichum oblongifolium* produces violet, lavender, or occasionally pink flowers to 2 cm (0.75 in.) across, plant height 102 cm (40 in.). 'Raydon's Favourite', a selection of the species, has violet flowers to 2.5 cm (1 in.) across, and grows more than 100 cm (40 in.) tall, while another selection, 'October Skies', has purple-blue flowers to 2 cm (0.75 in.) across, 45 cm (18 in.) in height.

Symphyotrichum 'Ochtendgloren'

Michaelmas daisy
Synonym *Aster* 'Ochtendgloren'

With generous sprays packed with purple-pink flowers, 'Ochtendgloren' is one of the best varieties selected for garden planting by Piet Oudolf, chosen from plants originally bred specifically to provide cut flowers. The most distinctive feature is its reliable performance in borders or containers even when left undivided for a number of years.

ZONES 4–8

HABIT AND SIZE The sprays are upright and sturdy to 120 cm (4 ft.) tall once established. Numerous open branches up the length of the sprays are borne from strong but compact clumps.

FLOWERS 2.5 cm (1 in.) across, in midautumn.

CULTIVATION Full sun, in good garden soil, which does not become waterlogged over winter. Space 60 cm (24 in.) apart. Divide every three to five years. Mildew resistant.

ORIGIN 'Ochtendgloren' like 'Anja's Choice' is another hybrid raised by Piet Oudolf in the Netherlands.

LANDSCAPE AND DESIGN USES This cultivar makes an excellent plant for the border, whether it is formal or more naturalistic, and a superb container plant. In a container, it will happily flower down the entire height of the spray provided it has been given enough feed and water during the summer.

SIMILAR PLANTS *Symphyotrichum* 'Phoebe' and *S.* 'Pink Star' appear to be identical in all respects to 'Ochtendgloren' and to one other.

Symphyotrichum 'Oktoberlicht'

Michaelmas daisy
SYNONYMS *Aster* 'Oktoberlicht', *A.* 'Octoberlight'

This is probably the best white-coloured aster for the dual-purpose use of a cut flower and garden plant. The masses of pure white flowers have relatively small pale yellow discs that turn brown with age.

ZONES 4–8
HABIT AND SIZE The clumps are strong but compact, producing sturdy, upright, well-branched sprays to 102 cm (40 in.) tall when established.
FLOWERS 2.5 cm (1 in.) across, in midautumn.
CULTIVATION Full sun, in good garden soil, which does not become waterlogged over winter. Space 60 cm (24 in.) apart. Divide every three to five years. Mildew resistant.
ORIGIN Unknown

LANDSCAPE AND DESIGN USES When grown in groups in a border, 'Oktoberlicht' can be used in the front as well as the centre to give a lift to the bed. This aster also helps provide a textural break whether it is among other small-flowered varieties, in which case its flowers appear much larger, or among other autumn herbaceous plants where it provides the airy side of the textural bargain. The compact growth and habit of flowering right down the stem makes it well suited to container culture.
SIMILAR PLANTS *Symphyotrichum* ×*salignum* 'Caledonia' produces masses of white flowers on sturdy growth like 'Oktoberlicht', but it can be invasive, 102 cm (40 in.) tall.

Symphyotrichum 'Photograph'

Michaelmas daisy
SYNONYM *Aster* 'Photograph'

Somehow, this small-flowered lavender-blue aster always seems to stand out in the garden and is desired by all those who see it. Despite its modest constitution and unfortunate tendency to be a feast for slugs in the spring, 'Photograph' has an edge on other small-flowered varieties. There is elegance to the wiry arching sprays, and clearness to the shade of lavender-blue, which can never quite be shown in images or adequately described.

ZONES 4–8
HABIT AND SIZE Wiry, arching sprays to 102 cm (40 in.) in height from small clumps.
FLOWERS 1.5 cm (0.5 in.) across, in midautumn.
CULTIVATION Full sun, in good garden soil, which does not become waterlogged over winter. Space 60 cm (24 in.) apart. Divide every three to five years. Mildew resistant.
ORIGIN Raised in the United Kingdom before 1920.
LANDSCAPE AND DESIGN USES This cultivar is best in an uncrowded position in the border or container where it can have the little bit extra attention needed to flourish.
SIMILAR PLANTS The small flowers of S. 'Hon. Vicary Gibbs' are a similar shade of lavender-blue to 'Photograph' but the growth is more upright and less elegant, 150 cm (59 in.) tall. *Symphyotrichum* 'Ringdove' has small pale lavender flowers on upright sprays, 102 cm (40 in.) tall.

Symphyotrichum pilosum var. *pringlei* 'Monte Cassino'

Pringle's aster, September flower
SYNONYM *Aster pringlei* 'Monte Cassino'

Very much the iconic florist's aster, 'Monte Cassino' produces distinctive small white flowers with yellow discs floating on very fine stems among narrow ferny foliage.

ZONES 5–8

HABIT AND SIZE Small woody clumps bear slender stems to 90 cm (36 in.) tall. The flowers are carried in broad panicles.

FLOWERS 2 cm (0.75 in.) across, often not opening until midautumn when grown in the garden.

CULTIVATION Full sun, in good light garden soil, excellent winter drainage. Space 60 cm (24 in.) apart. Divide every three to five years. Mildew resistant.

ORIGIN Raised by Fuss in 1983.

LANDSCAPE AND DESIGN USES This beautiful plant, one of the first asters used in the cut flower industry, is probably better suited to being grown undercover than in a typical English garden. In general, it is easier to manage as a container plant and puts on a long-lasting display.

SIMILAR PLANTS *Symphyotrichum ericoides* 'Constance' is a modern wiry cultivar with lovely clear white flowers excellent for cutting, on stems 102 cm (40 in.) tall.

Symphyotrichum 'Prairie Purple'

Michaelmas daisy
SYNONYM *Aster* 'Prairie Purple'

Huge numbers of small deep purple flowers start opening in early autumn and continue until midautumn. The leaves are dark and the stems are purple tinted, giving the whole plant a rich appearance throughout the year.

ZONES 4–8
HABIT AND SIZE Strong compact clumps produce sturdy stems to 150 cm (59 in.) tall. The sprays are well branched and dense.
FLOWERS 2.5 cm (1 in.) across, from early autumn.
CULTIVATION Full sun, in good garden soil. Space 60 cm (24 in.) apart. Divide every three to five years. Mildew resistant.
ORIGIN Paul Picton raised this aster in the early part of the twenty-first century. It is a cross between 'Little Carlow', the influence of which can be seen in the size, shape, and prolificacy of flowers, and 'Calliope', which has lent the dark colouration to stems and foliage.

LANDSCAPE AND DESIGN USES Despite the height of this cultivar it rarely needs any staking, particularly when grown hard (in low-nutrient, free draining conditions) in naturalistic planting schemes alongside perennial grasses. Ideal for the border where it provides colour over a long period during the latter part of the season. The warmth of the colour works well among other autumnal shades, particularly yellow. A personal favourite combination is the giant Joe Pye weed, *Eutrochium purpureum*, planted behind this cultivar where the warm purple-pink melds with the dense rich purple of 'Prairie Purple'.
SIMILAR PLANTS *Symphyotrichum* 'Little Carlow' resembles 'Prairie Purple' but with lavender-blue flowers borne over a shorter period, reaching a height of 120 cm (4 ft.).

Symphyotrichum turbinellum

Smooth violet prairie aster

SYNONYM *Aster turbinellus*

This distinctive aster has blue-green clasping foliage along upright arching sprays. Individual lavender-blue flowers are borne from long side branches of the sprays. The overall effect is light and airy, the blueness of the foliage becoming more apparent in drier conditions.

ZONES 4–8

HABIT AND SIZE Compact clumps produce long arching sprays to 120 cm (4 ft.).

FLOWERS 2.5 cm (1 in.) across, in midautumn.

CULTIVATION Full sun, in any reasonable garden soil. Space up to 90 cm (36 in.) apart. Divide every three to five years. Mildew free.

ORIGIN Grows wild on dry soils in Illinois to Missouri, Nebraska, Kansas, Louisiana, and Arkansas. Whether the plant I have described is a true wild species is the subject of some debate, but for now we are satisfied that it is.

LANDSCAPE AND DESIGN USES Super as a specimen plant, in a mixed border, or among grasses. As a specimen plant or mixed with a lightweight grass the stems can really arch, often with a diameter of 120 cm (4 ft.) from a tight clump. Even if you don't have that sort of space, it still looks attractive among other herbaceous plants in a traditional border. Surprisingly for a tall aster, it can make an excellent container-grown plant as well, but again needs a fair amount of room.

SIMILAR PLANTS *Symphyotrichum* 'Diamond Jubilee' inherited the same graceful growth as *S. turbinellum* but has white flowers slightly tinged with the palest violet. *Symphyotrichum* 'Turbinellus Hybrid' has very wiry dark stems with larger and slightly darker flowers than the species; it tends to lose its leaves before flowering. Both grow to 120 cm (4 ft.) tall.

GROWING AND PROPAGATING

Overall, asters are quite straightforward to grow in the garden and to propagate, if a few basic needs are met. As with many large and varied plant groups, it is difficult to apply a single set of rules for all. This chapter is designed to help you to decide what position to plant your asters in, and when and how to propagate them.

Site and Soil

A good rule for asters is that they need an open sunny position to thrive. There are of course some exceptions, such as white wood aster (*Eurybia divaricata*), which flourishes in a more shaded position. In the plant entries, we have outlined each cultivar's requirements for position and soil. Your soil will play an important role in choosing which plants you are going to grow.

If you have naturally free-draining soil, for example, it is advisable to avoid *Symphyotrichum novi-belgii* cultivars as they are shallow rooting and need plenty of water during summer months to grow and flower well. Good moisture levels will also help to reduce disease problems. Plenty of asters will thrive on free-draining soil, although many, particularly *S. novae-angliae* cultivars, will be shorter than when grown in more moisture-retentive soil.

Similarly, if you garden on a poorly nourished soil you might choose to avoid the greedy species such as *Symphyotrichum novi-belgii*. Instead, *Aster amellus* and many others will do quite well for you. It is worth remembering that almost all plants grown for excellent floral displays will do better if provided with some food.

Like all the New York asters, late-flowering *Symphyotrichum novi-belgii* 'Blue Gown' grows best in a moisture-retentive garden soil.

Cold-Hardy Asters

Aster alpinus
Aster amellus cultivars
Aster ageratoides 'Starshine'
Aster peduncularis
Aster trinervius var. *harae*
Doellingeria umbellata
Eurybia sibirica

Planting

In general, having chosen the spot to plant your asters, it is important to undertake some soil preparation to give them a good start. Firstly, the soil should be recently dug, which will make it easier for you to plant. This will also reduce compaction, which helps the plants' roots grow out and speeds the rate of establishment, thus decreasing the risk of water stress.

It is also a good idea to feed asters when planting. In our garden we use organic poultry manure pellets spread over the prepared soil (the exact application rates are given when you purchase the fertilizer of your choice). This method works well if you have a large area to plant, but if you were only planting an individual plant or single group of plants, it would be better to dig the hole and then mix a feed into the soil at the base before planting. Don't let the roots come into direct contact with the fertilizer as this can damage them.

A substantial clump of bright lavender-blue *Symphyotrichum* 'Little Carlow' thrives in the same well-drained soil that suits a pink-flowered sedum.

With its compact, self-supporting habit and outstanding flowers, *Aster ×frikartii* 'Wunder von Stäfa' makes an ideal specimen plant in the garden.

If you are planting *Aster amellus* or another species that needs good winter drainage into heavy soil it is advisable to mix grit into the base of the hole before planting to improve drainage.

Be sure to water plants immediately after planting and kept them moist until they are established.

SPACING

When choosing sites for asters it is important to remember that they start growing later in the spring than the earlier flowering herbaceous plants and therefore need to have space around them so that they do not get overcrowded. A common reason for people losing asters is that they have planted them in a mixed border where vigorous early summer herbaceous growth has smothered their young shoots. This not only cuts out vital sunlight but also provides a fantastic home for slugs and snails.

When planting a group of asters you need to consider two points: How big are the plants going to grow? And how long are they going to stay in the ground? Both the height and spread will determine how far apart they should be planted. If you are planning on lifting and dividing annually, as might be the case with *Symphyotrichum novi-belgii*, then the plants are not going to need as much space between them as they would if they were going to be there for two or more years. Again, you can find recommended planting distances in each plant description, but here is a more general guide:

If plants are 10–40 cm (4–16 in.) tall, then space them 30–45 cm (12–18 in.) apart.

If plants are 40–90 cm (16–36 in.) tall, then space them 30–60 cm (12–24 in.) apart.

If plants are more than 90 cm (36 in.) tall, then space them 40–60 cm (16–24 in.) apart.

Asters for Shade

Eurybia divaricata 'Eastern Star'
Eurybia ×*herveyi* 'Twilight'

Along with size, the vigour of the particular plant and the growth habit will also affect the spacing requirements. For *Symphyotrichum novi-belgii,* you can simply use the upper end of the general spacing guide for vigorous cultivars and the lower end for less vigorous ones. In the case of *S. novae-angliae,* it is necessary to use the upper end of the spacing guide, 60 cm (24 in.) apart, when planting in a group as they make very woody clumps that will spread over the years, and they should be left undivided for at least three years.

Plants of *Aster amellus* can usually be given the wider spacing for their height because of their bushy habits. The opposite is found with a number of the small-flowered asters such as *Symphyotrichum* 'Prairie Purple' which are best with a slightly reduced spacing for their height, maybe 45 cm (18 in.) apart.

Enough space should be left between groups to prevent neighbouring plants from invading each other over winter. This is particularly true with *Symphyotrichum novi-belgii* cultivars where rogue shoots from a neighbouring plant can lead to the wrong plant being propagated in early spring when it is hard to tell the varieties apart.

WHEN TO PLANT

The ideal time for planting most container-grown asters is in spring, after the ground has begun to warm up and before the weather becomes too hot and dry. Another option, especially for tough varieties, is to plant in late autumn when the ground is still warm but the weather is cooler. With European asters, such as *Aster amellus,* it is usually best to plant in late spring, as they are dormant over winter and so planting in late autumn does not give them time to establish a good root system, thus increasing the risk of losing them over winter. In our garden, the optimum time for planting out is late spring, but this is mainly because we use young plants divided in the early spring or late winter and so they need time to form a good root system before being committed to the open garden.

Spring- or Summer-Blooming Asters

Aster alpinus
Aster tongolensis 'Napsbury'

Large and spreading, *Symphyotrichum* 'Les Moutiers' produces billowing masses of dancing pink stars on dark stems. As the flowers age, the yellow disc florets turn purple.

Aftercare

As soon as your asters have been planted and watered in, you need to think about whether they will need support for their mature growth or not. Generally, this will depend on height, with most asters over 90 cm (36 in.) tall needing support of some kind. Suitably sized bamboo canes can be used to support individual sprays and small clumps. With careful placing, they become "invisible" when the plants are in full growth. If available free of charge, sticks cut from hazel or similar coppiced trees might be superior to bamboo. In the good old days, when professional gardeners were in abundant supply, they often wove beautifully designed supports of hazel twigs to cover entire herbaceous borders or specialist aster borders. In early spring, these borders looked like works of art!

A raised bed at Brobury Gardens in Herefordshire features asters of various heights, including some that need staking.

You can also choose from a wealth of metal supports, but beware that many are both expensive and ugly, and some are useless as well. Look carefully for practical, unobtrusive designs tall enough to support the asters at flowering time. This last point is particularly important since a support that is too short can almost be worse than no support as the stems break where the support ends if there is too much height above it.

Symphyotrichum novae-angliae cultivars should not need staking after their first year, as the woody base and stems are normally sufficiently strong to support the flowering stems. Similarly, other asters with a very bushy growth will probably not require staking just as those in a central position in the border may not. It is always worth staking *S. novi-belgii* cultivars over 60 cm (24 in.) tall as it is so disappointing when an autumn storm comes through and you find all your beautiful flowers on the ground.

The "Chelsea chop" as it is known in the United Kingdom (or the "aster whack" in the United States) is something we are often asked about. Both terms refer to the process of cutting the plant back to about 15 cm (6 in.) above ground level in late spring, the same time as the Royal Horticultural Society Chelsea Flower Show. We do not recommend this technique simply because of the extremely variable results between different cultivars and species, as well as year to year. However, it is popular and can in some regions and under certain conditions (such as in some parts of the United States where the growing season is shorter and faster than the United Kingdom) produce some excellent results including shorter growth with very tall varieties, bushier growth, and often later flowering.

We experimented with this process in our garden and found that the only asters amenable to it were the *Symphyotrichum novi-belgii* varieties. The *S. novae-angliae* cultivars survived but produced a mass of weak shoots from

Award-Winning Plants

THE FOLLOWING PLANTS are deemed particularly good performers in the garden by the Royal Horticultural Society and have been awarded the RHS Award of Garden Merit.

Pink

Aster alpinus
Symphyotrichum 'Coombe Fishacre'
Symphyotrichum ericoides 'Pink Cloud'
Symphyotrichum novae-angliae
   'Andenken an Alma Potschke'
Symphyotrichum novae-angliae 'Harrington's Pink'
Symphyotrichum novi-belgii 'Fellowship'
Symphyotrichum 'Ochtendgloren'

Lavender-blue

Aster alpinus
Aster ×frikartii 'Wunder von Stäfa'
Symphyotrichum cordifolium 'Chieftain'
Symphyotrichum 'Little Carlow'
Symphyotrichum 'Photograph'

Purple

Aster alpinus
Aster amellus 'King George'
Aster amellus 'Veilchenkönigin'

White

Symphyotrichum ericoides 'Golden Spray'
Symphyotrichum ericoides f. *prostratum*
   'Snow Flurry'

the top of the cut stems, which mostly failed to flower at all and any flowers produced were of a poor quality. Some *S. novi-belgii* cultivars, such as 'Mount Everest', worked really well one year and terribly the next. Our main advice is that if you want to try it, then do; if you are unsure, perhaps try just half a clump and if it works, try the entire clump the following year.

Once asters are established, they are unlikely to need watering, unless conditions have caused them to become exceptionally dry. In the first year, it is advisable to keep an eye on moisture levels. Particularly in late summer when most varieties are fattening their flower buds, it is worth watering if the soil is drying out and is likely to remain dry for some time. This is especially the case if the plant is showing signs of stress, such as the upper leaves beginning to droop or take on a greyish hue rather than their normal healthy green colour. Be slightly more vigilant with *Symphyotrichum novi-belgii* cultivars because not allowing them to get too dry will help alleviate the problem of powdery mildew. You can also help prevent excessive loss of moisture from the soil by mulching in the early spring or after planting. As an additional benefit for older clumps of asters, the right sort of mulch can also provide a certain amount of feed.

Asters are surprisingly greedy plants. Although feeding them is not strictly necessary for their survival, they will show a significant difference in the quantity and quality of bloom if you add a top dressing of a complete plant food (such as fish, blood, and bone) or use a proprietary slow-release feed. Additional feeding is not usually required unless your soil is very impoverished, or you are growing some specimen blooms for a flower show. In these circumstances, it would be advisable to make use of liquid plant food, especially in the early stages of flower bud development.

Once asters are established, they are unlikely to need watering, unless conditions have caused them to become exceptionally dry. In the first year, it is advisable to keep an eye on moisture levels. Particularly in late summer when most varieties are fattening their flower buds, it is worth watering if the soil is drying out and is likely to remain dry for some time.

Diseases and Pests

The diseases and pests that affect asters, while annoying, are largely superficial and can be easily remedied. The most important thing is to identify what is damaging your plant so you can decide what action to take. We have recommended some basic methods of control, and you may well find other options too.

POWDERY MILDEW

The main disease problem that has haunted asters over the years is the notorious powdery mildew. This fungal disease affects a range of common edible and ornamental plants including many of the Asteraceae (daisy family), apples, and marrows or squash. When a plant becomes infected, the fungal growth is superficial, covering just the surface of the plant. It can also affect the vigour of the plant and does look unsightly when a heavy infection occurs.

Like all *Symphyotrichum laeve* cultivars, 'Nightshade' is
virtually never troubled by powdery mildew, making it
a good trouble-free option for colour in midautumn.

Infection is often associated with plants suffering from water stress, which makes mulching to reduce water loss an important part of the natural control of mildew. Removing infected sprays as soon as the fungal growth is visible can also help to slow down the spread. Most strains of powdery mildew are specific to certain hosts so mixed planting can also decrease the risk of infection in the first place and reduce the rate of spread if infection does occur. If you are growing plants that are susceptible to mildew it is very important to clear away old foliage over winter before the new growth emerges.

Even when you've done all you can to reduce the risk of infection, it may still happen. In this case, you can investigate the numerous sprays, both organic and chemical fungicides, available on the market. No matter what you choose to use, sprays are normally most effective as a preventative measure rather than after infection has occurred.

Mildew-Free Asters

Aster ageratoides 'Starshine'
Aster alpinus
Aster amellus
Aster peduncularis
Aster pyrenaeus
Aster sedifolius
Aster tongolensis 'Napsbury'
Aster trinervius var. harae
Doellingeria umbellata
Eurybia ×herveyi
Eurybia sibirica
Symphyotrichum cordifolium 'Chieftain'
Symphyotrichum novae-angliae
Symphyotrichum oblongifolium 'Fanny's Aster'
Symphyotrichum turbinellum

The group most susceptible to powdery mildew is the New York asters (*Symphyotrichum novi-belgii*), although variation exists here as well. The other groups of asters are either completely free of mildew or resistant to infection. A large part of the resistance exhibited is physical: resistant plants have many more hairs on the surface of the foliage, which prevent the mildew spores from settling and infecting them. Those that are resistant will usually not be troubled by mildew except in exceptionally hard years where they suffer from water stress over a long period. Even then, the infection is usually mild and does not harm the plant, although infected material should still be removed at the end of the season.

VERTICILLIUM WILT

This soil-borne fungus infects the water-carrying tissues of the plant. In asters, the infection leads to wilting of the upper parts of the plant—the infected stems usually become discoloured at the base often followed by the foliage becoming brown. If your plant is infected it is important to remove it and as much of the root system as possible and destroy it. The soil will remain infected and will have to be treated or replanted with plants that are resistant to Verticillium wilt.

Some asters are more susceptible than others. *Aster amellus* can be affected where high levels of the fungus are present; it is best to prepare the ground with fresh soil if you are replanting after division in the same place. *Symphyotrichum novi-belgii* cultivars also tend to be susceptible to infection, the problem becoming notably worse when the plants have been replanted in the same place for a number of years. Changing where the divisions are planted will reduce the risk of this problem arising. If you are troubled about leaving the ground fallow for a while, sow a natural biofumigant green manure such as Caliente mustard. This beneficial blend of brown and white mustard seeds produces weed-smothering foliage, and when incorporated into the ground, improves soil fertility and structure and suppresses soilborne pests and diseases.

RUST

For once, this is a fungal disease that is not heading straight for *Symphyotrichum novi-belgii* cultivars and hybrids, but it can affect *S. novae-angliae* cultivars. The early symptoms are brown- or rust-coloured spots on the foliage and sometimes on stems or flowers. A severe infection can lead to the plant's death, but it is usually quite easy to control by removing infected leaves in the early stages. As with mildew, old plant material should be cleared away over winter if there has been a problem with rust. Rust is often carried on willow herb (*Epilobium angustifolium*) and groundsel (*Senecio vulgaris*), so it is worth ensuring that not too many of these are growing near your New England asters.

The diseases and pests that affect asters, while annoying, are largely superficial and can be easily remedied.

TARSONEMID MITES

These microscopic mites graze the growing tips of certain plants, famously strawberries. When this happens to asters, growth is distorted and, if left untreated, the plants often fail to flower or have very distorted flowers. The only treatment options are commercial sprays or the biological control option of using predatory mites. Luckily, tarsonemid mites mainly occur where plants are grown intensively so this is rarely a problem for home gardeners.

Autumn takes over the garden.

Cool-coloured New York asters combine well with the warm shades found in autumnal colour schemes.

APHIDS

Like tarsonemid mites, aphids are rarely a problem in the normal garden, but a very heavy infestation on young growth can cause distortion and a loss of vigour. It is worth keeping an eye on plants in the first stages of growth and looking for twisted young shoots, wrinkled young leaves, and a general lack of vigour. When any of these symptoms appear, a closer look will soon reveal numerous aphids tucked into the leaf axils and on the stems of the young shoots. If you do not wish to use a garden chemical, you will find that soapy water and some messy work with your fingers works well and is cheaper. Better still, if you live in an area where aphids abound each year, try preventative spraying with an organic plant oil–based product.

SLUGS

Slug damage seems to vary a lot between gardens but is rarely a major problem for asters. Damage is often seen on the young shoots of *Symphyotrichum novi-belgii* but they usually grow through it without any detrimental effect. At the nursery, we are not troubled by slugs on *Aster ×frikartii* or *A. amellus*, but over recent years, we have had reports from customers of serious damage inflicted on these plants. As with any plant that is favoured slug fodder, you will see evidence of grazing across the top of young shoots, and in the worst cases, this will mean that the shoots have been eaten off level with the ground. If this happens, you'll want to take action so slugs don't get the upper hand in your garden.

Along with garlic sprays, copper rings, eggshells, and beer traps, you can also find more eco-friendly compounds than the old standbys. The best solution we have found is a handful of call ducks—unlike chickens they do not damage your garden and they just love slugs. A gardener who lived near us invested in a good hand lamp and a large bucket. She went out every evening at dusk and the next day took her catch to an area of woodland about a mile away. But be warned: It is rumoured slugs have a very good homing instinct and will return, albeit slowly. Truth be told, ask ten gardeners how to get rid of slugs and you will get ten different answers. In reality, all you can do is manage the population if they are causing serious problems and be super vigilant around prime plant specimens.

A clump of single-shoot asters.

Propagation

The best way to propagate asters is by division of well-grown clumps, which should be done on a regular basis to keep the plants healthy and the flowers of a good size. The three methods of division correspond to the main groups of asters.

Once your aster is divided you can either replant the divisions if they are large enough, or grow them on in pots. Division should generally be done in the spring (early spring, just as new growth appears, is always a safe option), although if you live in an area with favourable winter conditions, it can also be done in autumn. Because our nursery is located on heavy, winter-wet land, we do all of our propagation in spring.

Division is usually the best method, but asters can also be propagated by top cuttings and by seed. Top cuttings are only beneficial if you want to produce large quantities of plant material. If you do want to try top cuttings, they can be taken from the spring growth of most asters, as soft wood shoots. These will root most readily in a mist propagation unit, but simple home-style propagators and cold frames in the garden will also work perfectly well when the essential extra time and effort is put into the job.

Propagation by seed is mostly very easy, however, plants will not come true from seed. You will usually end up with a mixture of hybrids, except for New England asters and *Aster amellus*, which will most likely be the right species but could be any colour or form. If you want your new plant to be identical to the named variety you already have, you must propagate vegetatively, either by division or by cuttings.

Separated from its clump, this shoot is ready to be replanted.

Newly planted shoots.

DIVISION OF SINGLE SHOOTS

This technique is suitable for New York asters (*Symphyotrichum novi-belgii*) and smooth aster (*S. laeve*) cultivars.

To start the process, lift a clump of plants and shake off the soil from the roots. Be careful not to break too many shoots.

Separate each shoot and cut it off from the clump with a sharp clean blade. Make sure you have approximately 2.5 cm (1 in.) of leaf and some roots on the shoot. Each shoot should be 5–7.5 cm (2–3 in.) long.

Place one to three shoots in each pot, depending on the size of pot and strength of the shoots. It is best to use a 7.5-cm (3-in.) pot for single shoots or a 9-cm (4-in.) pot for up to three shoots.

Fill your pot about a third of the way with a good quality potting compost, then place your shoots into the pot and fill with compost.

These pots of shoots must go undercover, such as in a well-lit porch, greenhouse, or polytunnel, for three to six weeks until roots can be seen emerging from the base of the pot. This is not to protect them from the cold but to stop the pots from becoming water-logged, which can lead to the shoots rotting off before they have made sufficient roots.

The new plants should then be hardened off before being planted out.

DIVISION OF WOODY CLUMPS

This technique is suitable for large, woody, clump-forming asters such as the New England asters (*Symphyotrichum novae-angliae*), or older clumps of other strong asters.

Start by lifting the clump and knocking off as much soil as possible. Use a spade or two forks to break the clump apart. Create sections of plant material, approximately 15 cm (6 in.) in size. These can then be set out for replanting once the ground has been prepared.

A clump of plants with excess soil removed.

The same clump divided into four smaller clumps for replanting.

DIVISION OF SMALL WOODY CLUMPS

This technique is suitable for European asters and the less vigorous small-flowered asters.

Lift the clump and shake off as much soil as possible. All the shoots will be grouped closely together in a tight crown.

Gently divide the crown into pieces with three or four flowering stem bases and a good number of new shoots. Use a small hand fork, sharp knife, or secateurs to achieve this without damaging too many roots or shoots. Do not attempt to break the clump down into sections with only one flowering stem base, as these will take a very long time to clump up even if they do survive.

Replant the divisions back into the ground, preferably in fresh soil. Alternatively, pot them up and grow them on until you are ready to plant them out.

Gently divide a tight crown of shoots into pieces using a sharp knife.

Each piece of the crown to be replanted has more than one flowering stem base.

WHERE TO BUY

Asters often appear in garden centres during the flowering season but caution should be taken when purchasing pots of brightly coloured "dwarf" asters: despite appearances, these have all too often been treated with growth-regulating hormones to keep them short. This makes it easier to handle them in the commercial world and easier to sell; however, the following year people are often horrified to find their dwarf aster now standing 90 cm (3 ft.) tall. It is worth making the effort to find a reliable nursery or garden centre that really knows their plants and preferably has propagated them too. The following list is a selection of a few reliable nurseries that stock a good range of asters. Of course, there are many more than we have space to list so do check out places in your local area.

CANADA

Hortico
422 Concession 5 East
Waterdown, Ontario L0R 2H1
www.hortico.com
Mail-order only.

Mason House Gardens
3520 Durham Road #1, RR #4
Uxbridge, Ontario L9P 1R4
www.masonhousegardens.com

Veseys
411 York Road
Highway 25
York, Prince Edward Island C0A 1P0
www.veseys.com/ca/en

FRANCE

Pépinière Poiroux
Le Petit Beauregard
D760-85340 Olonne sur Mer
www.pepiniere-poiroux.fr

GERMANY

Arends and Maubach
Monscahustrasse 76
42369 Wuppertal
www.arends-maubach.de
The famous perennial breeder Georg Arends originally established this 125-year-old perennial nursery and garden.

Stauden Feldweber
Hermine Gruber A-4974
Ort im Innkreis 139
www.feldweber.com

Staudengärtnerei Gräfin von Zeppelin
Weinstrasse 2
D-79295 Sulzburg-Laufen/Baden
www.staudengaertnerei.com

Staudengärtnerei Schöllkopf
Inh. Michael Frank
Gewand Heckwiesen
72770 Reutlingen
www.staudengaertnerei-schoellkopf.de
Specialist growers of chrysanthemums, asters, and alpines.

Stauden-Stade
Beckenstrang 24
46325 Borken-Marpeck
www.stauden-stade.de

Werner Simon
Staudenweg 2
97828 Marktheidenfeld
www.gaertnerei-simon.de

ITALY

Priola Pier Luigi
Via del Acquette
Treviso
www.vivaipriola.it
Specializing in perennials, ferns, grasses, and shrubs with an impressive range of asters for sale.

NETHERLANDS

Nursery 'de Hessenhof'
Miranda and Hans Kramer
Hessenweg 41
6718 TC Ede
www.hessenhof.nl

UNITED KINGDOM

Avondale Nursery
Mill Hill
Baginton, Near Coventry
West Midlands
England CV8 3AG
www.avondalenursery.co.uk
Nursery offering a wide range of perennials. Holder of three National Plant Collections of Aster novae-angliae, Sanguisorba, and Anemone nemorosa.

Beth Chatto Gardens
Elmstead Market
Colchester
Essex
England CO7 7DB
www.bethchatto.co.uk
Nursery and garden.

The Botanic Nursery
Atworth
Wiltshire
England SN12 8NU
www.thebotanicnursery.co.uk

Cotswold Garden Flowers
Sands Lane
Badsey
Evesham
Worcestershire
England WR11 7EZ
www.cgf.net
An English nursery specializing in unusual perennials and shrubs, owned by the well-known Bob Brown.

Elizabeth MacGregor
Ellenbank
Tongland Road
Kirkcudbright
Dumfries and Galloway
Scotland DG6 4UU
www.elizabethmacgregornursery.co.uk

Great Dixter House and Gardens
Northiam
Rye
East Sussex
England TN31 6PH
www.greatdixter.co.uk
Beautiful Arts and Craft house and garden with a nursery stocking many of the plants found in the garden.

Hardy's Cottage Garden Plants
Priory Lane Nursery
Freefolk Priors
Whitchurch
Hampshire
England RG28 7NJ
www.hardys-plants.co.uk
Extensive range of herbaceous plants.

Hoo House Nursery
Gloucester Road
Tewkesbury
Gloucestershire
England GL20 7DA
www.hoohouse.plus.com
*Wholesale and retail perennial
plants and alpines.*

Larch Cottage Nurseries
Melkinthorpe
Penrith
Cumbria
England CA10 2DR
www.larchcottage.co.uk

Little Heath Farm Nursery
Little Heath Lane
Potten End
Berkhamsted
Hertfordshire
England HP4 2RY
www.littleheathfarmnursery.co.uk

Norwell Nurseries
Woodhouse Road
Newark
Nottinghamshire
England NG23 6JX
www.norwellnurseries.co.uk
*Specializing in rare and unusual
perennials, many of which are displayed
in the garden next to the nursery.*

The Nursery Further Afield
Evenley Road
Mixbury, Near Brackley
Northhamptonshire
England NN13 5YR
www.nurseryfurtherafield.co.uk

Old Court Nurseries
Walwyn Road
Colwall
Malvern
Worcestershire
England WR13 6QE
www.autumnasters.co.uk
*Specializing in autumn-flowering asters
and other perennials. Holder of the
National Collection of autumn-flowering
asters, which are displayed in the Picton
Garden adjacent to the nursery.*

Perhill Nurseries
Worcester Road
Great Witley
Worcestershire
England WR6 6JT
www.perhillplants.co.uk

Perry Hill Nurseries
Hartfield
East Sussex
England TN17 4JP
www.perryhillnurseries.co.uk

RHS Wisley Plant Centre
RHS Garden Wisley
Woking
Surrey
England GU23 6QB
www.rhs.org.uk/Gardens/Wisley
*Very extensive range of plants stocked
including many asters.*

Rougham Hall Nurseries
Ipswich Road
Bury St Edmunds
Suffolk
England IP30 9LZ
www.perennials-of-distinction.co.uk

Waterperry Gardens
Waterperry, Near Wheatley
Oxfordshire
England OX33 1JZ
www.waterperrygardens.co.uk
Gardens and plant nursery with
wide range of asters available.

Worlds End Garden Nursery
Mosely Road, Hallow
Worcestershire
England WR2 6NJ
www.worldsendgarden.co.uk

UNITED STATES

Arrowhead Alpines
PO Box 857
Fowlerville, Michigan 48836
www.arrowheadalpines.com

Bluestone Perennials
7211 Middle Ridge Rd
Madison, Ohio 44057
www.bluestoneperennials.com

Busse Gardens
5873 Oliver Avenue SW
Cokato, Minnesota 55321
www.bussegardens.com

Forest Farm
14643 Watergap Road
Williams, Oregon 97544
www.forestfarm.com

Goodness Grows
332 Elberton Road
PO Box 311
Lexington, Georgia 30648
www.goodnessgrows.com

Joy Creek Nursery
20300 NW Watson Road
Scappoose, Oregon 97056
www.joycreek.com

Lazy S's Farm Nursery
2360 Spotswood Trail
Barboursville, Virginia 22923
www.lazyssfarm.com/index.html

Plant Delights Nursery
9241 Saul's Road
Raleigh, North Carolina 27603
www.plantdelights.com

Prairie Nursery
PO Box 306
Westfield, Wisconsin 53964
www.prairienursery.com

Siskiyou Rare Plant Nursery
2115 Talent Avenue
Talent, Oregon 97540
www.siskiyourareplantnursery.com

Sunlight Gardens
174 Golden Lane
Andersonville, Tennessee 37705
www.sunlightgardens.com

Tripple Brook Farm
37 Middle Road
Southampton, Massachusetts 01073
www.tripplebrookfarm.com

WHERE TO SEE

Brobury House and Gardens
Brobury
Herefordshire
England HR3 6BS
www.broburyhouse.co.uk
A garden offering year-round interest
with a unique border of asters.

Durmast House
Bennetts Lane
Burley
New Forest
Hampshire
England BH24 4AT
www.durmasthouse.co.uk
The garden of this house was designed
by Gertrude Jekyll in 1907 and offers a
rare peek at Jekyll's designs, including
a double Michaelmas daisy border.

National Collection of
Symphyotrichum novae-angliae
Brockamin
Old Hills
Callow End
Worcestershire
England WR2 4TQ
+ 44 (0)1905 830370
This collection is curated by Margaret
Stone set in the 0.6-hectare (1.5-acre)
garden of her home Brockamin. Best visited
from mid-September to early October.

National Collection of
Symphyotrichum novae-angliae
Mill Hill
Baginton, Near Coventry
West Midlands
England CV8 3AG
www.avondalenursery.co.uk
This collection is housed at Avondale
Nurseries run by Brian Ellis. Best visited
from mid-September to early October.

National Collection of
Symphyotrichum novi-belgii
Temple Newsam House and Estate
Temple Newsam Road, off Selby Road
Leeds
West Yorkshire
England LS15 0AE
Run by Leeds City Council this
collection is in the walled garden
of the Temple Newsam Estate.

National Collection of *Aster amellus,*
Symphyotrichum ericoides, **and**
S. cordifolium
Upton House and Gardens
Near Banbury
Warwickshire
England OX15 6HT
www.nationaltrust.org.uk/upton-house
A lovely collection set in the gardens
of the National Trust property Upton
House at Banbury, Oxfordshire. Best
visited from August to late October.

**National Collection of
Autumn-Flowering Asters**
The Picton Garden
Old Court Nurseries
Walwyn Road
Colwall
Malvern
Worcestershire
England WR13 6QE
www.autumnasters.co.uk
*Collection of many different species that
flower from midsummer onwards.*

Waterperry Gardens
Waterperry, Near Wheatley
Oxfordshire
England OX33 1JZ
www.waterperrygardens.co.uk
*A beautiful 3.2-hectare (8-acre) garden
with asters in traditional herbaceous
borders and stock beds.*

UNITED STATES

Bellevue Botanic Garden
12001 Main Street
Bellevue, Washington 98005
www.ci.bellevue.wa.us/botanical_garden.
htm

Chicago Botanic Garden
1000 Lake Cook Road
Glencoe, Illinois 60022
www.chicagobotanic.org

Longwood Gardens
1001 Longwood Road
Kennett Square, Pennsylvania 19348
www.longwoodgardens.com

Missouri Botanical Garden
4344 Shaw Boulevard
St. Louis, Missouri 63110
www.missouribotanicalgarden.org

New York Botanical Garden
2900 Southern Boulevard
Bronx, New York 10458
www.nybg.org

Wave Hill
West 249th Street and Independence
Avenue
Bronx, New York 10471
www.wavehill.org

FOR MORE INFORMATION

BOOKS

Brickell, Christopher, ed. 2008. *The Royal Horticultural Society A–Z Encyclopedia of Garden Plants*. 3rd ed. London: Dorling Kindersley.

Cubey, Janet, ed. 2014. *The RHS Plant Finder*. London: Dorling Kindersley.

Flora of North America Editorial Committee, eds. 2006. *Flora of North America*, vols. 19–21. New York: Oxford University Press.

Picton, Paul. 1991. *The Gardener's Guide to Growing Asters*. Devon: David and Charles; Portland, Oregon: Timber Press.

Ranson, E. R. 1946. *Michaelmas Daisies and Other Garden Asters*. London: John Gifford.

Robinson, William. 1907. *The English Flower Garden*. 10th ed. London: John Murray.

Schöllkopf, W. 1995. *Astern*. Stuttgart: Ulmer.

WEBSITES

Encyclopaedia of Life: www.eol.org

The Royal Horticultural Society: www.rhsplants.co.uk

USDA Plants: www.plants.usda.gov

ORGANIZATIONS

Chicago Botanic Garden: A Comparative Study of Cultivated Asters, www.chicago-botanic.org/downloads/planteval_notes/no36_asters.pdf

Mt. Cuba Center: Research Report—Asters for the [U.S.] Mid-Atlantic Region, www.mtcubacenter.org/horticultural-research/trial-garden-research

HARDINESS ZONE TEMPERATURES

USDA ZONES & CORRESPONDING TEMPERATURES

Temp °F			Zone	Temp °C		
−60	to	−55	1a	−51	to	−48
−55	to	−50	1b	−48	to	−46
−50	to	−45	2a	−46	to	−43
−45	to	−40	2b	−43	to	−40
−40	to	−35	3a	−40	to	−37
−35	to	−30	3b	−37	to	−34
−30	to	−25	4a	−34	to	−32
−25	to	−20	4b	−32	to	−29
−20	to	−15	5a	−29	to	−26
−15	to	−10	5b	−26	to	−23
−10	to	−5	6a	−23	to	−21
−5	to	0	6b	−21	to	−18
0	to	5	7a	−18	to	−15
5	to	10	7b	−15	to	−12
10	to	15	8a	−12	to	−9
15	to	20	8b	−9	to	−7
20	to	25	9a	−7	to	−4
25	to	30	9b	−4	to	−1
30	to	35	10a	−1	to	2
35	to	40	10b	2	to	4
40	to	45	11a	4	to	7
45	to	50	11b	7	to	10
50	to	55	12a	10	to	13
55	to	60	12b	13	to	16
60	to	65	13a	16	to	18
65	to	70	13b	18	to	21

FIND HARDINESS MAPS ON THE INTERNET.

United States *http://www.usna.usda.gov/Hardzone/ushzmap.html*
Canada *http://www.planthardiness.gc.ca/*
Europe *http://www.gardenweb.com/zones/europe/* or *http://www.uk.garden web.com/forums/zones/hze.html*

ACKNOWLEDGEMENTS

We would both like to thank our families and friends for their patience and help both on and off the nursery during the process of writing this book. We are also extremely grateful to the owners of the beautiful gardens where many of the images were taken, including Mrs. Pru Cartwright of Brobury House, Mr. John Massey VMH of Ashwood Nurseries, Mr. and Mrs. Cole of Meadow Farm, the Marquess and Marchioness of Hertford of Ragley Hall, and Olive Mason of Dial Park.

PHOTO CREDITS

Photographs are by Paul Picton unless indicated otherwise.

ROSS BARBOUR, pages 90, 108, 112, and 212.
ANABELLE DE CHAZAL, pages 59 bottom and 86.

INDEX

ABOUT THE AUTHORS

PAUL PICTON and his daughter, **HELEN**, are specialist growers of autumn-flowering asters. Along with Paul's wife, Meriel, and Helen's husband, Ross Barbour, the family operates Old Court Nurseries and the Picton Garden in Herefordshire, England. The nursery was established in 1906 by Ernest Ballard, the first nurseryman to popularize autumn-flowering asters. Since then it has always specialized in breeding and growing these super plants.

Every autumn several thousand visitors come to the nursery and garden to see the asters at their peak. The Pictons hold the Plant Heritage National Plant Collection of autumn-flowering asters, with more than four hundred forms present in the collection. It is a work of love and passion to keep the collection up together and make the garden attractive and ever changing.

Paul started working alongside his father, Percy, at Old Court when he was sixteen and has been surrounded by asters his entire life. With more than fifty years of experience, he has a vast, irreplaceable knowledge of the genus. He has travelled all over the country to give horticultural talks and is well respected within the horticultural community. Paul is author of *The Gardener's Guide to Growing Asters*.

Helen joined the business after completing a botany degree at the University of Reading, and has since then been working full time with the asters. She also lectures and gives talks throughout the United Kingdom.

Visit the family's nursery and garden at autumnasters.co.uk.

Front cover: Autumn colours
Spine: *Symphyotrichum novi-belgii* 'Lassie'
Title page: Colourful Asters
Contents page: *Symphyotrichum novi-belgii*

Published in 2015 by Timber Press, Inc.

The Haseltine Building   6a Lonsdale Road
133 S.W. Second Avenue, Suite 450   London NW6 6RD
Portland, Oregon 97204-3527

For details on other Timber Press books and to
sign up for our newsletters, please visit our websites,
timberpress.com and timberpress.co.uk.

Library of Congress Cataloging-in-Publication Data
Picton, Paul, 1942-
  The plant lover's guide to asters/Paul Picton and Helen Picton.—First
edition.
    pages cm
  Includes index.
  ISBN 978-1-60469-518-2
  1. Asters. I. Picton, Helen. II. Title.
  SB413.A7P54 2015
  635.9'3399—dc23
                                    2014024945

A catalogue record for this book is also available from the British Library.

Mention of trademark, proprietary product, or vendor does not constitute
a guarantee or warranty of the product by the publisher or author and does
not imply its approval to the exclusion of other products or vendors.

Book and cover design by Laken Wright
Layout and composition by Will Brown
Printed in China

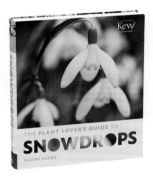